How To Build
Ford Flathead V-8 Horsepower

George McNicholl

How To Build
Ford Flathead V-8 Horsepower

George McNicholl

MOTORBOOKS

First published in 2005 by Motorbooks, an imprint of MBI Publishing Company, Galtier Plaza, Suite 200, 380 Jackson Street, St. Paul, MN 55101-3885 USA

Motorbooks titles are also available at discounts in bulk quantity for industrial or sales-promotional use. For details write to Special Sales Manager at MBI Publishing Company, Galtier Plaza, Suite 200, 380 Jackson Street, St. Paul, MN 55101-3885 USA.

ISBN-13: 978-0-7603-2290-1
ISBN-10: 0-7603-2290-2

Editorial: Lindsay Hitch and Peter Bodensteiner
Design: LeAnn Kuhlmann

Printed in China

On the front cover: Gary Scheibal owns this beautiful 1932 Ford Model B Roadster. The car was turned into a hot rod in 1953, and an Ardun engine was installed in 1987. Gary purchased this vehicle in 1991 and has spent the past 10 years restoring it to its original 1953 condition. The current Ardun 284-ci engine was installed in April 2003, and the buildup of this engine is featured in *How To Build a Flathead Ford V-8.* ***Front cover inset:*** Luke's Custom Machine & Design completed this Harrell blower intake manifold. The pop-off valve was fabricated using an overhead valve (OHV) Ford 2-inch-diameter intake valve.

On the title page: Tim Young owns this spectacular blown Ford flathead V-8 engine. The 265-ci engine is installed in the frame of his 1936 Ford five-window coupe. A Chevrolet S-10 pickup five-speed manual transmission is bolted to the engine. The Vintage Vendors exhaust headers are metallic-ceramic coated. The buildup of this engine is also featured in *How To Build a Flathead Ford V-8.*

On the back cover: Luke's Custom Machine & Design machined this polished-aluminum beehive oil filter, mounted on an Edelbrock 1115 cylinder head using a polished stainless-steel bracket.

About the Author

George McNicholl was born in Vancouver, B.C., Canada, more than 50 years ago. He purchased his first automobile engine (a Ford flathead V-8) when he was 12 years old. He sailed on deep-sea ships for 15 years, with the last two years in the position of captain. He is still involved in the shipping business, although he has been ashore for 25 years. Car and engine building have always been hobbies for McNicholl, and he has the greatest affection for the Ford flathead V-8. Suitably, he is the author of two other Motorbooks publications, *How To Build a Flathead Ford V-8* and *How To Build Vintage Hot Rod V-8 Engines.*

CONTENTS

FOREWORD

I have been an automotive enthusiast and devoted hot rodder for more than 40 years. The first book I wrote, *How to Build a Flathead Ford V-8*, was published by MBI Publishing Company and released at the end of April 2003. I shall not bring the reader to tears by rewriting my biography in this book. If anyone is interested, which is not likely, they may kindly purchase the first Ford flathead V-8 book and read about my background in the foreword of that publication. Do not think the sales of more of my books will result in any direct benefit to me—my family will just invent new ways to spend the revenue!

I have been compiling the complete buildup information for my own engines and vehicles for over 20 years, and for all the engines I have assisted Luke's Custom Machine & Design (see Appendix B: Resources) with assembling. That is where I spend any spare time I have, helping with engine assembly, detailing, and ordering parts. I assisted with the complete assembly of the three engines described in this book. At the end of each chapter, I have provided a summary section that contains a full description of the parts, part numbers, clearances, and prices. This data is extremely valuable when determining the appraised value of a vehicle for insurance purposes, and it serves as a quick-reference source.

The three engines described in the first flathead book were built using 1949–1953 Ford flathead V-8 engine blocks. Shortly after that book was released, a number of readers suggested I write about the complete buildup of an engine using the 1946–1948 Ford flathead V-8 Model 59 engine block. Although I prefer to use the later-model engine block for a street-performance flathead, many devout Ford owners and fans regard the Model 59 engine block as the only true hot rod flathead engine block. I hope the inclusion of this engine will place me back in their good graces!

I originally intended to write about the Model 59 engine and add that information to a future printing of the first flathead book. My editor, Peter Bodensteiner, informed me that adding the Model 59 engine to my first book would require almost the same amount of work as required to publish a new book. The idea of an entirely new book really started the wheels turning. . .

I shall make a short alteration of course to mention that I had the pleasure of meeting Max Ellis in August 2002 at Luke's Custom Machine & Design. In 1936, Max designed an intake manifold for GMC trucks that allowed for the fuel conversion from gasoline to butane. Earl Evans made the patterns for that intake manifold, and they were cast in Evans' garage. In 1938, Evans made some patterns for Chevrolet intake manifolds and cylinder heads based on Max's designs. These were cast and marketed under the name of "Evans-Ellis." In 1942, Max Ellis and Earl Evans went their friendly separate ways. Evans continued on with his hot rod line of intake manifolds (one is shown in this book) and cylinder heads, including the now-famous models for the Ford flathead V-8. Max formed the Ellis Manifold Corporation to supply truck manifolds. I attempted to photograph Max next to the Ardun engine described in the first flathead book. Unfortunately, and mysteriously, my camera had absorbed some water and refused to function—another historic moment down the tube!

Although there are three completely different engines described in this publication, there is good information throughout the book that is applicable to all the engines. I highly recommend a thorough reading of this book in order to be eligible for the daily afternoon quiz! Most of the material and the vast majority of photographs in this book are new and not a repeat of the material and photographs in the first flathead book. There are a few overlaps, which were unavoidable, because some information is common to all of the engines. This book, combined with my first flathead book, will provide the reader with all the information necessary to build a naturally aspirated or blown 1946–1953 Ford flathead V-8 engine and Ardun V-8 engine for street use.

Hemmings Motor News (www.hemmings.com) and eBay (www.ebay.com) are excellent sources for locating Ford flathead V-8 engine blocks and related parts. I have purchased numerous items from eBay without any problems; however, I only purchase from highly rated sellers.

It is my desire that every person reading this book will gain at least one piece of helpful information from it. This will have made my time and effort in the writing of this manuscript worthwhile. Good luck with your own flathead or Ardun project. Remember to drive safely and sanely!

—George McNicholl
December 2004

ACKNOWLEDGMENTS

I would like to thank my wife, Jillian, for her continuing patience with me and for the use of her camera (once again), which I used to take every picture in this book. I know that is going to cost me dearly! I would also like to thank my son, Tyler, and my daughter, Kristina, for their computer assistance, without which this book would have been written using a quill pen. I am fascinated that kids today think their parents are borderline morons when it comes to computer skills!

I would sincerely like to thank Peter Bodensteiner, my acquisitions editor with MBI Publishing Company, for his ongoing support of my projects. Through his superb efforts, my first book was published and has become a great success. Peter is a true gentleman and, as an editor, is an inspiration to any author. This is also going to cost me!

I wish to express my gratitude to E. Dean Butler for providing me with previously unpublished information about the Ardun V-8 and V-8-60 engines and Zora Arkus-Duntov.

I would also like to express my gratitude to: Gregg Suyenaga at the technical department of Edelbrock Corporation; Keith Lee, the owner of Jitney Auto Parts; and Bill Jenks, the legendary camshaft grinder at Mooneyes USA, for providing me with some of the valuable camshaft data in Appendix C: Vintage Camshaft Specifications.

INTRODUCTION

"Part A" of each chapter contains a detailed explanation of the engine parts, the machining required, and where to locate the parts. "Part B" of each chapter contains the summary of the engine parts and clearances along with prices. The prices I have listed are those in effect as of December 2004. They do not include any shipping and handling charges, and they are based on the assumption that the owner of the engine will purchase his own parts. Generally, if the engine building or machine shop purchases the parts, you can count on a 20 percent markup. This is a legitimate charge: the shops must account for the time it takes to order the parts, inspect the parts, and pay for them. It is not a common practice to include prices in most automotive publications, as prices can change quickly over a very short period of time. I have done so in order to provide the reader with as accurate an indication as possible as to the cost of building similar types of engines. The cost of building one of these engines is not low; however, the price is reasonable considering the quality of the finished product. I should point out that the cost of building a top-quality stock Ford flathead V-8 engine would most likely be comparable to the cost of building a street-performance engine. The prices for the machine work would be the same; leaving out a high-performance intake system, aftermarket cylinder heads, and an electronic ignition system would reduce costs.

Note: These prices are in U.S. dollars. The prices indicated may vary in different parts of North America and are subject to change without notice.

There is a world of difference between an engine building shop that routinely rebuilds the average automobile engine and a high-performance-engine building shop where strict tolerances are the rule. Do not try and cut corners—the result will be extreme disappointment and most likely a waste of some hard-earned money. Leave the work to a professional engine builder. There are many critical clearances that must be checked when assembling an engine that the average person would overlook.

Some excellent sources of information for Ford flathead V-8 engines are the master repair manuals and chassis parts catalogs published by Ford Motor Company. These books are becoming hard to locate, although companies such as Wilkinson's Automobilia (see Appendix B: Resources) do stock this type of publication. The Ford parts numbers in this book are those listed in the *Ford Motor Company of Canada Master Repair Manual*—engine including cooling and exhaust system, manual number SE 348-B, 10/53; section number 12; price: 35 cents. Don't you just love the price?!

In this book, Section I covers the building of a street-performance 1946–1948 Ford flathead V-8 Model 59 engine. Section II covers the building of a blown street-performance French flathead V-8 engine. Section III covers the building of a blown Ardun V-8 street-performance engine.

SECTION I: MODEL 59 STREET-PERFORMANCE ENGINE

CHAPTER 1
ENGINE BLOCK

The 1946–1948 Ford flathead V-8 Model 59 engine is a street-performance motor with a displacement of 263 ci. The engine was built by Luke's Custom Machine & Design and completed at the beginning of November 2004.

Part A
Purchase and Magnaflux

The 1946–1948 Ford and Mercury flathead V-8 Model 59 specifications are:

239.4-ci displacement
Stock bore: 3 3/16 inches
Stock stroke: 3 3/4 inches
Stock compression ratio: 6.75:1
100 horsepower at 3,800 rpm and 180 ft-lb torque at 2,000 rpm

Ford Motor Company's dynamometer test for this engine model was carried out on November 16, 1945, using snappy 71-octane gasoline!

Ford Motor Company referred to this engine model as an "L-head" design based on someone's observation that the combustion chamber resembled an upside-down "L." (I think it looks more like a "Y"!)

The 1946–1948 Ford or Mercury Model 59 engine blocks are the best pre-1949 flathead engine blocks. These were the last flathead engine blocks with the integral bell housing and the first pre-1949 engines with the large (2.138 inches to 2.139 inches) connecting rod journal diameter. The 1946–1948 engines also had the same 100-degree angle between the valves as the 1949–1953 Ford flathead V-8 engines. The Model 59 engine block is a desirable candidate for a street-performance engine because it can be safely bored to 3 5/16 inches (3.3125 inches) and in many instances to 3 3/8 inches (3.375 inches), unlike some of the pre-1946 engine blocks.

The number "59" is usually found at the rear of the engine block on the top of the bell housing or in the valve

This picture shows the number "59" in the center of the valve lifter gallery, which has been painted with Glyptal G-1228A medium grey gloss enamel.

Here is a 1946–1948 Ford flathead V-8 Model 59L engine block with the factory combustion chamber relief.

lifter gallery and might read: "59," "59A," "59AB," "59L," "59M," "59X," "59Y," or "59Z." The 59X, 59Y, and 59Z engine blocks were supposedly used in military vehicles. The Canadian Ford blocks sometimes had "C59" or "C59A" on the top of the bell housing, occasionally at the right (passenger's side) front of the deck surface and usually on the side of the engine block, just behind the left or right water pump. The Model 59L and 59Y engine blocks that I have encountered were factory relieved. There is a good possibility there are other less familiar industrial or military casting numbers. The Model 59 engine blocks also had two 1/4-inch NPT water jacket drain plugs at the bottom front behind the water pumps. There is one hole on each side of the bottom front of the integral bell housing to attach frame support rods.

Thoroughly inspect the Model 59 engine block for any cracks or damage. Do not be surprised if you find cracks around the cylinder head stud holes between the cylinders. This is quite common and is probably the result of over-torqued cylinder head studs or a lack of antifreeze during the winter months in colder climates. The cracks around the cylinder head stud holes can be repaired by an experienced machine shop. Examine all the cylinder head stud holes to ensure they have not been stripped or drilled off-center by some clown attempting to repair stripped threads. It is best to try to locate an engine block that is not cracked.

It is far more serious if cracks are found around the valve seats, and the cracks lead to the cylinder walls. If this is the case, decline the engine block and press on with the search for a decent one. Good used Ford flathead V-8 engine blocks are available; it only takes time and some patience to find one. Visually examine the sides of the engine block, the oil pan rails, the valve lifter bosses, and the cylinder walls. Pay particular attention to the area around the main bearing webs; this area was supposedly strengthened in the 1949–1953 flathead blocks. There does not appear to be any major difference between the main bearing webs in these two models of engine blocks, aside from a little less material around the rear main bearing web in the Model 59 engine block.

If you do not know what to look for when purchasing an engine block, take along a friend who does. The best approach is to have the engine block inspected at a local engine building shop. If the seller will not agree to this proposition, walk away and find another engine block with a reasonable seller. You can buy a complete engine or just

This beast in the bushes is a 1946–1948 Ford flathead V-8 Model 59Y engine block. This engine block was factory relieved. The "Y" is upside down.

This is a 1946–1948 Canadian Ford flathead V-8 Model C59A engine block. The C59A is on top of the bell housing.

the engine block. It is generally more economical to buy just an engine block, if the builder has a source for individual parts. There are many parts on a stock engine that will not be used when building a street-performance Ford flathead V-8 engine.

The maximum bore that a Ford flathead V-8 Model 59 engine block can safely sustain is 3 5/16 inches plus 0.030 inch (3.3425 inches). Some engine blocks have been bored to 3 3/8 inches (3.375 inches), providing there has been no core shift during the casting process; however, the cylinder walls become pretty thin with a 3 3/8-inch (3.375-inch) bore. Anyone wishing to bore a flathead engine block to 3 3/8 inches (3.375 inches) should have the cylinders sonic tested to ensure there is an ideal minimum of 0.200 inch material left after the cylinder-boring process.

Our engine block example was bored to 3 5/16 inches plus 0.030 inch (3.3425 inches). If you want the same engine, you will have to locate an engine block that has been previously bored to a maximum of 3 5/16 inches (3.3125 inches). If the engine block is bored to 3 5/16 inches + 0.030 inch (3.3425 inches), there might still be enough material left in the cylinder walls for the engine block to be bored one more time in the future.

When purchasing an engine block, it should have the main bearing caps attached. It is not a crisis if the main bearing caps are missing, it just means the engine block will have to be align honed (possibly align bored) in order to fit new main bearing caps. Consider this additional cost when purchasing an engine block without main bearing caps. The current estimated cost for a 1946–1948 Ford flathead V-8 Model 59 engine block in good condition is $275.

Have the engine block and the main bearing caps and bolts Magnafluxed, technically known as "magnet particle tested." This will be the final test to determine if you are the proud owner of one of Henry Ford's finest achievements. The estimated cost to Magnaflux an engine block and the main bearing caps and bolts is $50.

Cleaning

Remove any bolts and fittings, the camshaft bearings, and the rear upper oil seal retainer. The Ford 6043 oil gallery line plug in the Model 59 engine is located at the front of the engine block behind the camshaft timing gear. The plug is 1/8-inch NPT, and there is a tip on the end, usually referred to as a "restrictor-type" plug. The 1949–1953 Ford flathead V-8 engines utilized a 3/8-inch NPT plug. The Model 59 engine blocks also had a 1/4-inch NPT plug at the rear of the engine block above the oil pump idler gear cover gasket area. Remove those two oil gallery line plugs. After the engine block has been completely stripped down, have it hot tanked. Do this before the engine block is Magnafluxed to ensure there are no hidden cracks covered up with grease and grime. The estimated cost to hot tank an engine block is $50.

This is the Ford 6043 front oil gallery line plug for the Model 59 engine block. If this is missing or damaged, use a stainless-steel 1/8-inch NPT plug.

It is a common practice with overhead valve (OHV) engines to grind the valve lifter gallery smooth in order to assist with the oil return to the oil pan. This procedure is not necessary with a flathead engine block due to the high quality of the casting. The valve lifter gallery is actually quite smooth in a flathead engine block. You have just been saved from another thankless task.

Porting and Polishing

The most opportune time to port and polish and gasket match the intake and exhaust ports is after the engine block has been hot tanked. Spray or brush some blue machinist's dye around the intake and exhaust ports, position the intake and exhaust manifold gaskets that will be used with the engine, and scribe around the gaskets. Use a high-speed grinder with a carbide bit to gasket match the ports. The idea is not to lighten the engine block by 50 pounds, so do not go berserk with the grinder in your hands. The purpose is to remove just enough material to match the gasket openings and clean up any casting flash in the ports. Lightly grind the intake ports all the way to the valve seats; do not grind the valve seats. Lightly grind the exhaust ports as far up as possible. The optimum shape for the end exhaust ports on each side of the engine block is a "D" shape.

This 1948 Canadian Ford Model C59A flathead V-8 engine block has been fully ported and polished. Notice the "D" shape of the end exhaust ports and the detailing of the engine block.

This Crower Cams & Equipment Company 66575 drill jig is used to drill slots in the Ford flathead V-8 valve lifter bosses. The slots prevent the groove-lock solid adjustable valve lifters from rotating.

Detailing and Valve Lifter Bosses

After the engine block has been Redi-Stripped, use a high-speed grinder with small sanding discs to remove the rough areas from the exterior surface of the engine block. The entire surface should be as smooth as possible. The exterior surface of most flathead engine blocks is fairly decent, so this should not entail a lot of work.

Drill a 3/16-inch (0.1875-inch) hole in the center of each of the valve lifter bosses. Each hole should be approximately 1 inch from the top of the valve lifter boss. The purpose of these holes is to simplify the valve lifter adjustment. A 3/16-inch (0.1875-inch) drill bit (use the chuck end), or a punch, is inserted in the hole to prevent the valve lifter (tappet) from turning. The valve lifter is then easily adjusted using a thin tappet wrench. This method is far superior to smashing your knuckles with the Johnson adjustable lifter wrenches usually supplied with valve lifters, and it takes half the time. This system will not work if the adjustable valve lifters do not have common oval indents or slots in the sides.

Full-Flow Oiling

The one item that was missing from all the flathead engine blocks was a full-flow oil system through the oil filter. The full-flow oil system in a 1946–1948 Ford flathead V-8 Model 59 engine block is different from the one installed in a 1949–1953 Ford flathead V-8 engine block. In most Model 59 engine blocks, there are only two oil line fitting holes at the rear of the engine block and these are used for the pressure-out lines. A drain-back oil line fitting hole is achieved by drilling and tapping a 3/8-inch NPT hole in the top of the bell housing next to the pressure out holes.

"X" marks the spot! This is where the oil line fitting hole should be drilled and tapped for the drain back line when installing a full-flow oil system in a 1946–1948 Model 59 engine block.

Here are the oil line fittings for a full-flow oil system on a Model 59 engine block. The horizontal fitting on the left side is for the pressure out line, the vertical fitting in the center is for the oil pressure gauge, and the fitting on the 90-degree elbow is for the drain back line. The engine block has been painted with PPG DGHS 9000 black high-gloss polyurethane.

The center of the drain-back hole should be 1 3/4 inches from the center of the vertical pressure-out hole toward the centerline of the engine block. On some Model 59 engine blocks, there is a raised lump where the drain-back oil hole should be located.

The horizontal hole for an oil line fitting on the lower left side near the back of the engine block will be used for the pressure-out line and should be drilled and tapped using a 3/8-inch NPT tap. This hole leads to a horizontal crossover passage inside the engine block. Drill and tap the crossover passage using a 1/4-inch NPT tap and then install a stainless steel 1/4-inch NPT plug in the passage using Permatex aviation form-a-gasket. The vertical threaded hole above the horizontal hole is drilled and tapped using a 1/4-inch NPT tap. This hole will also be used as a pressure-out line for an oil pressure gauge fitting.

It has been reported by some knowledgeable parties that the installation of a full-flow oil system does not directly filter the oil for the number 3 main bearing and the number 4 and number 8 connecting rod bearings. Even if this were the case, at least 66 percent of the main bearing oil and 75 percent of the connecting rod bearing oil is directly filtered. That is 100 percent better than what was happening with the stock flathead oil filter system. Eventually, the filtered oil will return to the number 3 main bearing and the number 4 and number 8 connecting rod bearings. The installation of a full-flow oil system in a street-performance flathead will greatly assist in ensuring a long and happy bearing life.

Relieving the Engine Block

The matter of relieving the Ford flathead V-8 engine block has been a major subject of discussion for the past 50 years. Some engine builders are staunch advocates of this procedure, while other engine builders are totally against it. Some engine builders will point out that the Ford Motor Company factory relieved some of its engine blocks in the 1940s. While this is true, Ford Motor Company did not continue with that practice. I do not intend to put forth a great discussion about this matter; however, I shall make a few quick points. The air/fuel mixture flows over the top and around the valves, not straight down from them, in a flathead engine. Relieving the engine block will lower the compression ratio approximately 0.5 to 0.75. I have not heard of a situation in which lowering the compression ratio in a normally aspirated engine leads to an increase in power. A street-performance Ford flathead V-8 engine block does not need to be relieved. Save a lot of money and forget about relieving the engine block.

Heat Risers

Years ago, it was the practice to insert a penny in each heat riser passage in order to block them off. Leave the heat riser passages open for street use. (This might be a nonissue due to the fact that some aftermarket intake manifolds do not have heat riser openings.) The purpose of the heat riser passages is to warm up the intake manifold as fast as possible when the weather is cold. If the heat riser passages are blocked off and you live in a colder climate, your vehicle will run horribly until the engine reaches normal operating temperature.

Retap and Chamfer Bolt Holes

Run a tap through every bolt hole in the engine block to guarantee each and every thread is in perfect shape when it comes time to reassemble the engine. This procedure will also ensure accurate torque readings. Use a high-speed drill with a countersink bit and lightly chamfer all the cylinder head stud (or bolt) holes. Take it easy when carrying out this task—the idea is not to see how far you can push the countersink into the engine block. This process will greatly assist in preventing the cylinder head stud (or bolt) hole threads from being pulled when torqued. Countersink the cylinder head stud (or bolt) holes after the engine block has been parallel decked.

Redi-Strip Engine Block

Redi-Stripping is a nonacidic process that completely removes all foreign material from an engine block. It is amazing to see what has actually been living inside an engine block, particularly after 50 or so years. Redi-Stripping ensures that the water jackets in a Ford flathead V-8 engine block are absolutely clean and helps to prevent overheating. A reputable engine building or machine shop will usually insist that a flathead engine block is Redi-Stripped prior to performing any major work on it. The estimated cost to Redi-Strip (see Appendix B: Resources) an engine block is $170.

The 1946–1948 Ford flathead V-8 Model 59 engines came equipped with cylinder head studs. If these studs have been in the engine block for any length of time, leave them in place and send the engine block to be Redi-Stripped. The studs will be easier (without guarantee) to remove after the Redi-Strip process. This may limit the possibility of breaking the studs or damaging the block threads when they are finally removed.

Glyptal

Glyptal is the paint used on electric motors. The paint will withstand very high temperatures, and it is extremely durable. I prefer to purchase this paint in the spray can (rattle can) form for ease of application. As soon as the engine block returns from being Redi-Stripped, paint the valve lifter gallery and the front of the engine block (behind the camshaft gear) with Glyptal G-1228A medium grey gloss enamel. Glyptal is available from electrical repair outlets. Purchase 32 plastic plugs, 1-inch diameter, from a hardware store and insert them in the valve lifter bosses and valve guide bosses. Wrap masking tape around them if they don't quite fit. Alternatively, you may use a discarded set of valve lifters and valve guides. Insert a rubber 1 1/4-inch-diameter bathtub plug in the opening for the valve lifter gallery crankcase breather tube. Plug the other five holes in the valve lifter gallery, and mask off the outside of the engine block with newspaper. The Glyptal paint can now be applied.

Machining

It is extremely important to locate an automotive machine shop that specializes in high-performance engine building to ensure that close tolerances are adhered to. I have been using the services of High Performance Engines (see Appendix B: Resources) for years. Their work is impeccable and reasonably priced. It might not be convenient to have your machine work performed by that particular company, but it is important to locate a machine shop with similar high standards.

This set of Ross Racing Pistons is laid out for the boring operation. Each piston is measured with a micrometer, and the cylinders are bored accordingly to allow for 0.004-inch piston-to-cylinder bore clearance (naturally aspirated motor).

Cylinder Boring

The engine block described in this section of the book was bored and the cylinders honed and deglazed, resulting in a final bore of 3 5/16 inches plus 0.030 inch (3.3425 inches). The pistons should be purchased beforehand and delivered to the machine shop with the engine block. The machine shop will then measure each piston with a micrometer and bore the block accordingly to ensure there is 0.004-inch piston-to-bore clearance (after cylinder honing), measured below the bottom of the wrist pin perpendicular to the wrist pin. This is the piston clearance that Ross Racing Pistons recommends for its forged aluminum pistons when installed in a naturally aspirated motor. The engine block is usually bored to within 0.003 inch to 0.004 inch of the final bore and then honed and deglazed, resulting in the final cylinder bore.

OHV engines should be bored and the cylinders honed and deglazed using a torque plate. The torque plate simulates the distortion caused to the engine block when the cylinder heads are torqued. On a flathead engine block, the cylinder head bolts (or studs) are not evenly spaced around each cylinder, particularly where the valves are located. For this reason, and the fact that the cylinder bore is relatively small, the use of a torque plate is not mandatory. If a torque plate is used, the cylinder head studs or cylinder head bolts that will be installed in the engine block must be used with the torque plate. Even if these two types of fasteners were torqued to the same setting, the distortion to the engine block would be different for studs and bolts.

The main bearing caps for the Model 59 engine block must be installed and torqued to 80 ft-lb (the 1949–1953 Ford flathead V-8 main bearing caps are torqued to 105 ft-lb) during the cylinder boring operation. This procedure places stress on the bottom end of the engine block and prevents the cylinders from distorting after they have been bored, when the crankshaft is installed and torqued in place. The cylinder banks in a flathead engine are exactly 45 degrees in relation to one another. The estimated cost for cylinder boring, honing, and deglazing a flathead engine block is $130.

The cylinder boring is carried out using a Rottler Boring Bar Company Model F2B boring bar or similar machine. The engine block is set up in this machine, the centerline of the crankshaft is located using a dial indicator, and the engine block is then clamped into position for the actual boring operation.

Many flathead engine blocks in the 1950s and 1960s had cylinder sleeves installed when the motors were rebuilt by Ford dealers. Most of these cylinder sleeves were manufactured using thin-walled material and some were not properly installed. There is nothing wrong with using cylinder sleeves in an engine block, provided it is a quality cylinder sleeve and the installation is done professionally. The material used in the manufacture of cylinder sleeves today is superior to the material used for the cylinders in the flathead engine blocks. If an engine block is going to have the cylinders bored and a cylinder sleeve is required for one of the cylinders, a premium high-performance engine building shop will install the cylinder sleeve first and then perform the cylinder-boring operation. If this sequence is not carried out, the cylinder next to the sleeved cylinder will distort after the cylinder sleeve is installed. The estimated cost to sleeve one cylinder is $150, including the sleeve.

The final stage of the cylinder-boring operation is the cylinder honing and deglazing. This important step ensures the proper seating of the piston rings. The procedure is undertaken using a Sunnen automatic cylinder resizing machine, Model CK-10, or similar equipment. This machine removes the last 0.003 inch to 0.004 inch of material from the cylinder walls using Sunnen 600-series

This Rottler Boring Bar Company Model F2B boring bar is used to bore the cylinders in a Ford flathead V-8 engine block.

The Ertel Products Company (Indianapolis, Indiana) SL-107 dry cylinder sleeve is designed for a Ford flathead V-8 engine. This cylinder sleeve is designed for a 3 5/16-inch bore; the length is 7 7/16 inches with a wall thickness of 3/32 inch (0.09375 inch).

This Sunnen automatic cylinder resizing machine (Model CK-10) is used for the final cylinder honing and deglazing of this Ford flathead V-8 engine block.

280-grit stones (for cast-iron piston rings). The final bore and finish on the cylinder walls will be a cross-hatch pattern.

Align Hone

The centerline of the crankshaft is the point from which all critical measurements are taken. The centerline of the crankshaft is used to ensure the engine block main bearing housing bores, with the main bearing caps attached, are exactly parallel to it. The crankshaft centerline in the Ford flathead V-8 engine blocks is offset 0.265 inch to the right.

The engine block main bearing housing bores, with the main bearing caps attached, must be measured to ensure they are in perfect alignment with one another and they are not tapered. If this is not the case, the engine block will have to be align honed. Each time the align honing operation is carried out, approximately 0.0005 inch is removed from the main bearing housing bores and the main bearing caps. If the main bearing caps are not the original main bearing caps supplied with the engine block, align honing might be necessary, and the estimated cost for this is $200. If aftermarket main bearing caps are being installed, the engine block will most likely have to be align bored and align honed, and the estimated cost for both procedures is $360.

Install the main bearing caps in the engine block without the main bearings and torque the main bearing cap studs (or bolts) to 80 ft-lb (1949–1953: 105 ft-lb). Use an inside micrometer or a dial indicator to measure the inside diameter of the main bearing housing bores and main bearing caps. The 1946–1953 Ford flathead V-8 main bearing housing bore diameter is 2.670 inches to 2.671 inches; it should be perfectly round.

To check the main bearing housing bore alignment of the engine block use a crankshaft that has been checked for alignment and is guaranteed to be straight. The correct main bearings are installed in the engine block and the main bearing caps, the bearings are well oiled, the crankshaft is installed, and then the main bearing cap bolts (or studs) are correctly torqued. The crankshaft should spin freely without any tight spots. If it does, the engine block is satisfactory. If it does not, the engine block will have to be align honed.

The 1946–1948 Ford flathead V-8 Model 59 engine blocks came equipped with Ford 6345 main bearing cap studs and Ford 6348 main bearing cap stud nuts. The studs had a hole near the top that permitted wire to pass through the castle nut and the stud in order to prevent the nut from coming loose. It is not uncommon to find the studs have been replaced with cap screw bolts or a mixture of studs and bolts. This situation indicates that the engine block is most likely a candidate for an align-honing job. The engine block described in this section of the book had a mixture of studs and bolts. A decision was made to use all 1949–1950 Ford 7RA-6345-A cap screw bolts and not scrounge around for a few of the earlier style studs and nuts.

The distortion caused to the bottom end of an engine block is not the same when using studs and nuts as when using cap screw bolts, even when these two different types of fasteners are torqued to the same setting. The main bearing cap studs are the same diameter for their entire length, which will sometimes prevent a main bearing cap from positioning itself correctly if main bearing cap screw bolts have been previously used. If an engine block is align honed with cap screw bolts in place, you cannot replace

The book on the left is the Ford Motor Company of Canada Master Repair Manual, *engine including cooling and exhaust system, manual number SE 348-B. The book on the right is the* Ford Motor Company of Canada Chassis Parts Catalogue *for the 1949–1950 Ford and Meteor. Shop manuals such as these are available from Wilkinson's Automobilia.*

the bolts at any time thereafter with studs and nuts. The reverse situation also applies. It must be decided prior to align honing which type of fastener will be used.

It has been reported that the two cap screw bolts used for the front main bearing cap in the 1949–1953 Ford flathead V-8 engine were full shouldered, and these two bolts were installed without using lock washers. The bolts for the center main bearing cap and rear main bearing cap were not full shouldered, and they were installed using Ford 34849-S8 lock washers. The *Ford Motor Company of Canada Master Repair Manual* lists the same 6345 part number for all the main bearing cap bolts without any mention of full-shouldered bolts. The *Ford Motor Company of Canada Chassis Parts Catalogue* for the 1949–1950 Ford specifies the use of six lock washers. Whether the bolts are full shouldered or not, new aircraft-quality lock washers should be installed with all the main bearing cap bolts when building a street-performance flathead engine.

The stock main bearing caps and studs (or bolts) in a Model 59 engine used for daily driving are of sufficient

The main bearing caps on the left are Ford 6329 (F), 6330 (C), and GAU-12420 (R) for the 1946–1948 Model 59 engine blocks. The main bearing caps on the right are Ford 6323 (F), 6330 (C), and 6325 (R) for the 1949–1953 flathead blocks. These main bearing caps have been shot peened to remove stress.

strength to withstand normal stress in the bottom end. If the engine is going to be highly modified for street use, main bearing cap support bridges and aircraft-quality 1/2-inch-diameter main bearing cap bolts should be installed on the number 1 and number 2 main bearing caps. The oil pan does not usually have to be modified to clear the main bearing cap support bridges. If a Model 59 engine is going to be equipped with a blower, new number 1 and number 2 main bearing caps will have to be fabricated and fitted, and aircraft-quality 1/2-inch-diameter main bearing cap bolts installed.

If the engine is going to be used for all-out racing, an oil pan rail and main bearing cap support girdle should also be fabricated and fitted to reduce high-speed stress. The use of new main bearing caps and/or a support girdle will require align boring and honing and modification to the oil pan for main bearing cap clearance.

The main bearing caps supplied with the Model 59 engine blocks did not use as much material in their manufacture as the 1949–1953 models, and some of the early caps did not have an "F" for front or a "C" for center cast on the top. The Model 59 front cap is Ford 6329, and the center cap is Ford 6330. The front cap has less material across the front face of it than the center cap. The rear main bearing cap is Ford GAU-12420.

A close examination of the bearing surface of the main bearing caps for the engine described in this chapter revealed some shiny streaks around them, indicating that the main bearings were being pounded. Between that and the mixture of the studs and bolts, this engine block definitely needed to be align honed.

Parallel Deck

An engine block is parallel decked to ensure the deck height is the same for all the cylinders. It also guarantees the deck is exactly 90 degrees in relation to the cylinder bores. Most flathead engine blocks require the removal of 0.005 to 0.012 inch of deck material when they are parallel decked. The estimated cost to parallel deck a Ford flathead V-8 engine block is $120.

A Repco Automotive Equipment Company Type ASG or similar machine is used to parallel deck the engine block. The crankshaft centerline is located using a dial indicator. The engine block is clamped into position, and then one side of the engine block is parallel decked, followed by the other side.

This Repco Automotive Equipment Company Type ASG parallel decking machine is used to parallel deck a Ford flathead V-8 engine block. Notice the valve lifter adjustment holes in the valve lifter bosses.

Shot Peen

While the engine block is at the machine shop, have the main bearing caps and bolts shot peened to remove stress from those items.

Valve Seat Inserts

If the original intake and exhaust valve seat inserts are badly pitted or cracked, they should be replaced. The engine described here will be fitted with 1.61-inch head diameter intake and exhaust valves. There is enough material left around the original intake valve seat inserts when a 1.61-inch head diameter intake valve is installed. The intake valve seat inserts run cooler than the exhaust valve seat inserts, so they do not need as much material around them. There is enough material left around the original exhaust valve seat inserts to install 1.61-inch head diameter exhaust valves.

A majority of engine builders recommend installing hardened exhaust valve seat inserts for use with today's unleaded gasoline. Many Ford flathead V-8 engine blocks were fitted with hardened exhaust valve seat inserts at the factory. A machine shop can tell immediately if there are hardened exhaust valve seat inserts installed when the valve seat inserts are being ground. If a vehicle equipped with a flathead engine will not be driven on a daily basis, hardened exhaust valve seat inserts are really not required. The engine block described in this chapter already had hardened exhaust valve seat inserts. The intake and exhaust valve seat inserts were in very good condition and did not require replacement.

If new valve seat inserts are required for the installation of 1.61-inch head diameter valves, install Ertel Manufacturing Company 80780 intake valve seat inserts and 80810 hardened ductile iron exhaust valve seat inserts. The valve seat inserts can be purchased from most automotive supply stores or through your machine shop. Have an experienced machine shop install the valve seat inserts. The estimated cost to install 16 new valve seat inserts is $200 (not including the cost of the valve seat inserts).

Valve Grind

Locate a machine shop that has a reputation for performing top-quality Serdi-machined multiangle blueprint valve grinds. The Serdi machine is equipped with a boring bar that machines the valve seat inserts, rather than the old method of grinding the valve seat inserts with stones. The boring bar blends the valve seat inserts in with the intake and exhaust port runners, also referred to as pocket porting. A Serdi-machined valve grind ensures all the valves are perfectly centered in the valve seat inserts and they are all at the same depth. A Serdi-machined multiangle blueprint valve grind was performed on the intake and exhaust valve seat inserts in the Model 59 engine block. The estimated cost for a Serdi-machined multiangle blueprint valve grind on a Ford flathead V-8 engine block is $225. High Performance Engines knocks down the Serdi machine in order to

The Ford flathead V-8 engine block is being attacked by a Serdi 60 (automatic centering by spherical and flat air cushion) valve grinding machine. This is the ultimate in valve grinding and the only way to grind valve seat inserts.

perform a multiangle blueprint valve grind on a Ford flathead V-8 engine block. Bud and Dave Child have designed two jigs for centering purposes. This permits the valve seat inserts in a flathead engine block to be ground with or without the valve guides installed.

A Serdi 60 (automatic centering by spherical and flat air cushion) valve seat grinding machine should be used. The Serdi machine performs outstanding valve seat grinding and is the best way to grind the valve seat inserts in a Ford flathead V-8 engine.

Deglaze

Deglaze the valve lifter bosses and the valve guide bosses using a ball hone or a wheel cylinder hone. Do just enough honing to clean up the bosses; remove only an absolute minimum of material to prevent excess valve lifter and valve guide clearance, which would result in oiling problems.

Cracks

If there are any cracks around the cylinder head stud (or bolt) holes, have them stitched by an expert machine shop. This is absolutely not a job for the home engine builder or even many welders. There is a procedure to repair the cracks in an engine block that must be followed in order to avoid completely destroying the engine block.

Cleaning and Painting

After all the machining procedures described in this chapter have been performed, thoroughly clean the engine block with hot water and soap. Use long thin engine brushes to reach as far into the oil passages as possible. Spend the time to do this job properly—the term "surgically clean" applies here. Rinse the engine block off with hot water and use compressed air to blow out the engine block until it is dry.

Mask off the engine block with newspaper and use the intake manifold gasket, cylinder head gaskets, oil pan gaskets, water pump gaskets, oil pump idler gear cover gasket, and timing gear cover gasket as templates for the masking. This way all the exposed areas of the engine block will be painted. The intake manifold gasket should be trimmed to the shape of the intake manifold. This is referred to as detailing.

Polyurethane paint is without question the best paint to apply to an engine block. This paint is impervious to lacquer thinner, brake fluid, antifreeze, and gasoline, and the finish is outstanding. PPG is one of the best suppliers of polyurethane paint, available in just about every color imaginable. PPG polyurethane paint can be purchased from most industrial paint supply stores or even body shops.

Part B

- 1948 Ford 239.4-ci L-head V-8 Model 59 engine block; casting number: 12661; two-bolt main bearing caps; 24-stud block; stock bore: 3 3/16 inches (3.1875 inches); stock stroke: 3 3/4 inches; stock compression ratio: 6.75:1; 100 horsepower at 3,800 rpm and 180 ft-lb torque at 2,000 rpm; engine block weight (bare): 190 pounds; **$375**
- Engine block hot tanked and Redi-Stripped; engine block, main bearing caps, and main bearing cap bolts Magnafluxed; **$245**
- All threads retapped; main bearing caps and main bearing cap bolts shot peened; cylinder head bolt holes chamfered; valve lifter bosses deglazed, allowing for 0.002-inch valve lifter clearance; valve lifter bosses drilled with 3/16-inch (0.1875-inch) hole for valve lifter adjustment; new Weatherhead 31524-4 polished brass water jacket drain plugs, 1/4-inch NPT, installed using pipe thread sealant; exterior surface of engine block sanded and detailed; valve lifter gallery and front of engine block painted with Glyptal G-1228A medium grey gloss enamel; exterior surface of engine block painted with PPG DGHS 9000 black high-gloss polyurethane; **$595.90**
- Engine block converted to full-flow oil system; **$55**
- Engine block bored 0.155-inch (1/8 inch plus 0.030 inch) oversize and cylinders honed and deglazed using Sunnen 600-series 280-grit stones; final bore: 3.3425 inches (3 5/16 inches plus 0.030 inch); piston-to-bore clearance: 0.004 inch measured below bottom of wrist pin perpendicular to wrist pin; **$130**
- Engine block align honed; **$200**
- Engine block parallel decked to 0.010-inch average, below deck; **$120**
- Serdi-machined multiangle blueprint valve grind performed; **$225**
- Intake and exhaust ports fully ported and polished and gasket matched; **$216.67**

Engine Block Total: $2,162.57

CHAPTER 2
CRANKSHAFT

Part A
Purchase

The best crankshaft for a street-performance Model 59 engine is the 1949–1953 Mercury flathead V-8 crankshaft with the 4.00-inch stroke. A Mercury crankshaft was not available for the engine described in this book, and as a result, a 1949–1953 Ford 6303 crankshaft with the 3 3/4-inch (3.75-inch) stroke was installed in the engine to accommodate the late-model connecting rods. The Ford crankshaft is every bit as good as the Mercury crankshaft; both were manufactured using cast alloy steel. The *Ford Motor Company of Canada Master Repair Manual* specifies the out-of-round service limit at 0.0015 inch, the taper service limit at 0.001 inch, and the maximum runout at 0.002 inch. The Ford crankshafts can still be purchased at approximately half the cost, or less, of a Mercury crankshaft.

When purchasing a 1946–1948 Ford or Mercury flathead V-8 crankshaft, remember that the Model 59 main bearings are only available up to 0.030-inch oversize and the full-floating connecting rod bearings are only available up to 0.040-inch oversize. The main bearings for a 1949–1953 Ford or Mercury flathead V-8 crankshaft are only available up to 0.030-inch oversize and the nonfloating connecting rod bearings are available up to 0.060-inch oversize. If you need to grind the crankshaft in excess of those sizes, bearings are not currently available. The only solution to this problem is to have the journals built up. (This process is explained in this chapter.)

Because it is becoming harder to locate a Mercury crankshaft in decent condition, a flathead owner might want to consider purchasing a brand-new crankshaft from SCAT Enterprises. Their crankshafts are manufactured in the United States using 9,000-series material, which is said to be the strongest cast material available. Strokes of 4.00 inches, 4.125 inches (4 1/8 inches), 4.250 inches (4 1/4 inches), and 4.375 inches (4 3/8 inches) are available. The purchaser has the choice of the stock 2.138- to 2.139-inch connecting rod journal diameter, with or without a 1/8-inch (0.125-inch) billet radius on the journals, or a 2.00-inch connecting rod journal diameter for use with the SCAT Enterprises 2-239-7000-2000 H-beam

Here are two vintage crankshaft journal gauges. The Federal Mogul 2 (the large one) measures journals from 2.470 to 3.250 inches, and the Federal Mogul 1 (the small one) measures journals from 1.970 to 2.550 inches.

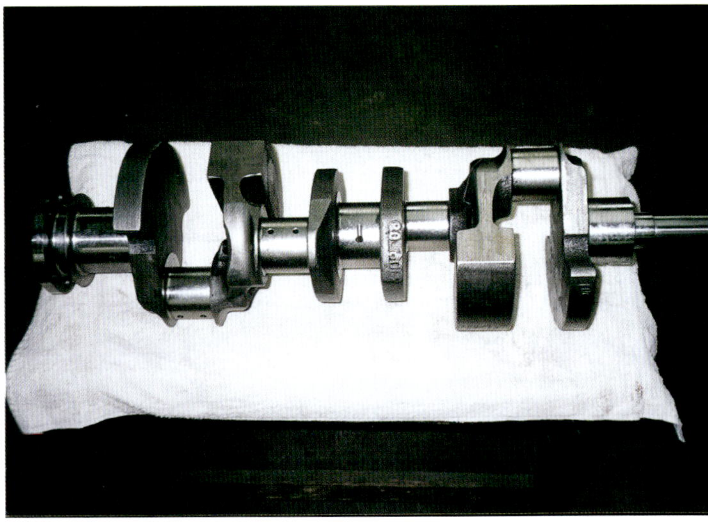

Here is a SCAT Enterprises 9-239-4250-2000 cast 9000 material crankshaft with a 4.25-inch stroke—high-quality crankshaft at an affordable price.

connecting rods. The main journals are the stock 2.498- to 2.499-inch diameter. These crankshafts are more than capable of handling even the wildest of street flathead engines. The SCAT Enterprise's crankshafts retail at $995—about half of what a custom billet steel crankshaft would cost. Just imagine all those cubic inches!

Magnaflux

Have an agreement in place with the seller when purchasing a used crankshaft. The purchase should be subject to the crankshaft being hot tanked, Magnafluxed, and thoroughly inspected. If the seller will not agree to this proposition, walk away; there will always be another crankshaft and a more reasonable seller. The estimated cost to Magnaflux a crankshaft is $25.

Shot Peen and Plug Removal

After the crankshaft has been hot tanked and Magnafluxed, it should be shot peened to assist removing any stress. Remove the four pressed-in plugs from the ends of the connecting rod journals. These plugs may have to be drilled out. The holes should be tapped with a 1/8-inch NPT tap (some crankshafts require 1/4-inch NPT or 3/8-inch NPT plugs). Then install stainless-steel pipe plugs using Loctite. Install the connecting rod journal oil plugs after all the machining procedures have been carried out and the crankshaft is thoroughly cleaned. Remove the crankshaft gear with a gear puller.

Straighten, Chamfer, and Polish

Check the crankshaft for alignment and runout with a dial indicator, and have it straightened if necessary. All the oil holes should be chamfered by lightly grinding the sharp surface off around the oil holes. The estimated cost to chamfer the oil holes is $25.

After the crankshaft has been ground and the oil holes have been chamfered, the journals should be polished to ensure an absolutely smooth bearing surface. The estimated cost to polish the crankshaft journals is $40.

Regrinding and Cleaning

If the crankshaft journals are standard size and it is necessary to grind the journals, grind them to 0.010-inch undersize. If the journals have been ground 0.010-inch undersize, grind them to 0.020-inch undersize, and so on. A quality machine shop will ensure the crankshaft is ground and polished to allow for the correct bearing clearance. The estimated cost to grind a crankshaft is $140.

A Mercury 4.00-inch-stroke crankshaft was brought to Luke's Custom Machine & Design last year. The owner of the crankshaft dearly wanted to salvage it, although one connecting rod journal was 0.065-inch undersize with the rest of connecting rod journals 0.020-inch undersize. There are two methods to build up connecting rod journals. The first is accomplished using wire welding to build up the journal, after which the journal is ground in the usual fashion. This is a proven method of repair that a quality crankshaft grinding company can easily handle. The estimated cost for wire welding each connecting rod journal is $140.

The second method of repair consists of spraying molten metal on the connecting rod journal, after which the journal is ground in the usual fashion. The connecting rod journal is then surface nitrided (case hardened) to 60

The Ford 6303 crankshaft has been ground, the oil holes chamfered, and the journals polished. It is now ready for the Model 59 engine block.

This 1949–1953 Mercury 4-inch-stroke crankshaft has had the two connecting rod journals built up using wire welding. This quality repair job is virtually undetectable after the crankshaft is ground.

Rockwell C hardness. A few months back, I had the opportunity to examine an early Chrysler Hemi V-8 crankshaft that had all the connecting rod journals sprayed in order to accommodate standard-size connecting rod bearings. The connecting rod journals appeared to have a very porous surface, which may have been the result of an inexperienced party performing the work. After viewing this type of repair, I recommend exercising caution when considering this method. The estimated cost to metal spray each connecting rod journal is $140.

Main Bearings

I have been using Clevite 77 (Dana Corporation) bearings in my own engines for the past several years. They are excellent bearings and competitive in price. The Clevite 77 bearings are a tri-metal construction with a soft lead alloy outer layer, a copper-lead center, and a steel backing.

At present, the Clevite 77 main bearings are not manufactured for the 1946–1948 Ford flathead V-8 Model 59 engines. The Federal Mogul 533M main bearings are currently offered in standard size, 0.010-inch oversize, 0.020-inch oversize, and 0.030-inch oversize for the 1946–1948 Ford flathead V-8 Model 59 engines. The Federal Mogul standard-type bearings are a tri-metal construction with a steel backing, a copper-lead center, a thin nickel barrier, and lead-tin-copper outer layer. The current retail price for a set of Federal Mogul main bearings for the Model 59 engine is $120. Main bearings in sizes other than those listed above might be available from Egge Parts House or Jitney Auto Parts (see Appendix B: Resources).

This set of Federal Mogul 533M-20 main bearings (0.020-inch oversize) was installed in the 1946–1948 Ford flathead V-8 Model 59 engine.

In order to check the main bearing clearance, insert the main bearings in the engine block main bearing bores and main bearing caps, install the main bearing caps, and then torque the main bearing cap bolts (or studs) to 80 ft-lb (105 ft-lb for a 1949–1953 flathead engine block). Use an inside micrometer or a dial indicator to measure the inside diameter of the main bearings. Use a micrometer to measure the main bearing journals on the crankshaft. The difference between these two measurements is the main bearing clearance. The *Ford Motor Company of Canada Master Repair Manual* specifies the main bearing clearance to be 0.000 to 0.0026 inch.

Check the crankshaft end play by inserting the main bearings in the engine block main bearing bores and the main bearing caps. Ensure that the main bearings are well oiled, the crankshaft is installed, and the main bearing cap studs (or bolts) are correctly torqued. Attach a dial indicator to the front of the crankshaft snout or to the end of the crankshaft flywheel flange. Insert a socket extension bar into one of the crankshaft counterweight balancing holes, and move the crankshaft forward and backward. The reading on the dial indicator is the crankshaft end play. The *Ford Motor Company of Canada Master Repair Manual* specifies the crankshaft end play to be 0.002 to 0.006 inch. The rear main bearing is the thrust bearing, which controls the end play.

Seldom does a Ford flathead V-8 engine block fitted with a Ford or Mercury crankshaft have a problem with crankshaft end play. These engine blocks and crankshafts are made of sturdy material. As long as quality main bearings are installed with the correct clearance, the crankshaft is perfectly aligned, and the original main bearing caps are in place, there should not be a problem with the crankshaft end play.

Have the machine shop that grinds the crankshaft supply the crankshaft bearings to be certain that you are in possession of the correct bearings and not something that fits the Wright Brothers' airplane engine. Federal Mogul and Clevite 77 bearings are available at most automotive parts supply houses.

Crankshaft Gear and Pulley

The stock 1946–1948 Ford 6306 cast-iron crankshaft gear is entirely suitable for use with a street-performance Model 59 engine. If this gear has been damaged or you are not able to locate one, the Melling 2704 is the replacement model available at automotive parts outlets.

We installed a 1940 Ford 6312 single-groove crankshaft pulley with the Ford 6332 crankshaft pulley bolt washer and the Ford 6319 crankshaft pulley bolt on the Model 59 engine. Exactly 1 inch must be removed from the length of the shaft at the rear of the crankshaft pulley if using a Luke's Custom Machine & Design crankshaft sleeve in order to correctly line up the V-belt. The outside diameter is 5 inches, and the 1940 Ford 8602 fan is bolted to the front of the crankshaft pulley. The six-blade fan will clear a crab-style distributor when mounted in this position, unlike the later-model flathead engines where the fan is mounted on the front of the generator or generator mounting bracket. The 1940 Ford 6312 crankshaft pulley requires the wide (5/8 inch) V-belt.

The 1946–1948 Ford 6319 crankshaft pulley bolt has four claws on it and was referred to as the starting crank ratchet. This bolt is designed for use with a hand crank to start the motor!

Part B

- 1949–1953 Ford 6303 cast alloy steel crankshaft; casting number: G81; main journal diameter: 2.498 to 2.499 inches; connecting rod journal diameter: 2.138 to 2.139 inches; nonfloating connecting rod bearings; stock stroke: 3 3/4 inches (3.75 inches); crankshaft weight: 67 pounds; **$100**
- Crankshaft Magnafluxed, aligned, and shot peened; main journals ground 0.020-inch undersize; connecting rod journals ground 0.020-inch undersize; oil holes chamfered; journals polished; crankshaft balanced; connecting rod journal oil plugs removed and new stainless-steel 1/8-inch NPT plugs installed using Loctite; **$270**
- New Federal Mogul 533M-20 main bearings, 0.020-inch oversize, installed allowing for 0.002-inch crankshaft clearance and 0.004-inch end play; 1949–1950 Ford 7RA-6345-A main bearing cap bolts torqued to 80 ft-lb using molykote; **$120**
- New Melling 2704 cast-iron helical crankshaft gear (22 teeth) and Ford 6310 oil slinger installed with Ford 357654-S crankshaft gear woodruff key; New Luke's Custom Machine & Design crankshaft oil seal sleeve installed; 1940 Ford 6312 crankshaft pulley (5-inch diameter, single groove) installed with Ford 74153-S crankshaft pulley woodruff key, Ford 6332 crankshaft pulley bolt washer, and Ford 6319 crankshaft pulley bolt; crankshaft pulley bolt installed using Loctite and torqued to 50 ft-lb; 1940 Ford 8602 six-blade fan installed with new stainless-steel hex head bolts and lock washers using Loctite and torqued to 25 ft-lb; oil slinger painted with Glyptal G-1228A medium grey gloss enamel; crankshaft pulley and fan painted with Endura EX-2C black 160 high-gloss poly-urethane; **$155.87**

Crankshaft Total: $645.87

This 1940 Ford 6312 single-groove crankshaft pulley and 1940 Ford 8602 fan were installed on the Model 59 engine.

CHAPTER 3
CONNECTING RODS AND PISTONS

Part A
Purchase

The 1946–1948 Ford 6200 (Model 29A) vanadium steel drop forged connecting rod has an average weight of 493 grams and is the full-floating type, used with large-journal (2.138 to 2.139 inches) crankshafts. There are no bearing tang (tab) notches in the Model 29A connecting rods, and they are 7.00 inches long (center to center). The Model 29A connecting rods have a hole near the end of the connecting rod studs that permits wire to be passed through the castle nuts and the studs to prevent the nuts from coming loose. These connecting rods can be used with the 1949–1953 Ford or Mercury nonfloating crankshafts, provided the thicker 1946–1948 Ford full-floating connecting rod bearings are used with them. Full-floating connecting rod bearings for the 1946–1948 flatheads are becoming hard to locate and are more expensive than nonfloating connecting rod bearings. The full-floating connecting rod bearings require more labor to achieve clearance than the nonfloating connecting rod bearings.

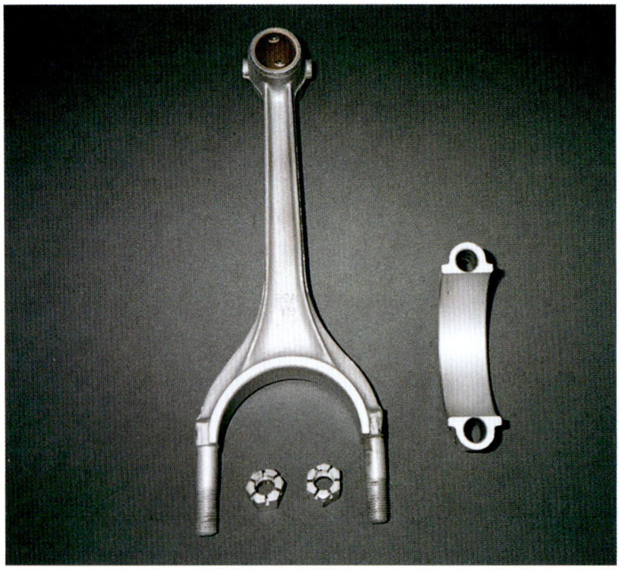

There are no bearing notches in the big end (crankshaft end) of the 1946–1948 Model 29A connecting rod used with the large journal (2.138 to 2.139 inches) crankshafts and full-floating connecting rod bearings.

Note: In my seemingly endless quest to provide you, the reader, with the maximum amount of information, I should note that there is another Model 29A connecting rod with a casting number of 29A6205 for use with the thicker 1946–1948 Ford full-floating, large-journal (2.138 to 2.139 inches) connecting rod bearings. The 29A6205 connecting rods have no bearing notches, but they have safety-wire holes in the studs and weigh 490 grams each.

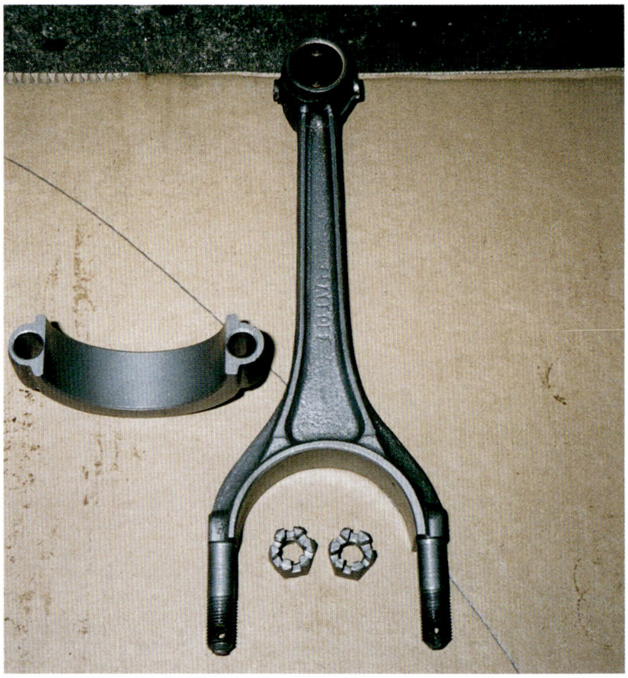

The big end (crankshaft end) of this 1946–1948 Ford flathead V-8 connecting rod (casting number 29A6205) does not have any bearing notches. This connecting rod is used with the large journal (2.138 to 2.139 inches) crankshaft.

Just to add a little more confusion to the issue, I shall mention that a very interesting set of connecting rods was discovered at Luke's Custom Machine & Design. These Ford flathead V-8 connecting rods are ex-Canadian military stock C49B-6200 and the casting number is also 29A6205. The casting number would indicate these connecting rods are for use with full-floating connecting rod bearings. However,

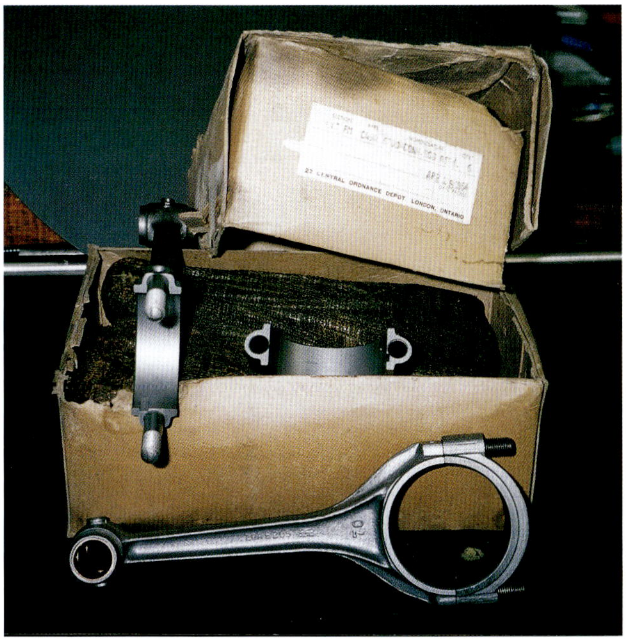

This set of Canadian Ford military C49B-6200 (casting number 29A6205) connecting rods is designed for use with the large journal (2.138 to 2.139 inches) crankshaft. These connecting rods have bearing notches, although only full-floating connecting rod bearings are used with these connecting rods. The box contains a set of six new old stock (NOS) C49B-6200 connecting rods wrapped in burlap and Cosmoline, dated April 28, 1954.

these connecting rods have bearing notches that were never used with full-floating bearings, and the nonfloating connecting rod bearings are not thick enough to use with these connecting rods. The bearing notch depth is shallower than the bearing notch depth in the 1949–1953 Ford 8BA6205A or 8BA connecting rods. The big end (crankshaft end) diameter of the 29A6205 connecting rods is 2.3595 to 2.360 inches—the same big end (crankshaft end) diameter of the 1946–1948 Ford Model 29A connecting rods and 0.069 inch larger than the 2.2905-to-2.2910-inch big end (crankshaft end) diameter of the 1949–1953 Ford 8BA6205A or 8BA connecting rods. Some of the Canadian Ford Military C49B-6200 connecting rod studs have holes near the end of the studs for safety wire and others do not. The average weight of a C49B-6200 connecting rod is 502 grams.

It is pure speculation on my part, but it would appear that the Ford Motor Company of Canada, at the request of the Canadian military, manufactured the C49B-6200 connecting rods during the same period that the 1949–1953 Ford 8BA6205A connecting rods were being produced. The big end (crankshaft end) of the C49B-6200 connecting rod was enlarged, permitting these connecting rods to serve as replacements for the 1946–1948 Model 29A connecting rods. This is one of the world's great unsolved mysteries and another good reason to use the 1949–1953 Ford flathead V-8 connecting rods, casting number 8BA6205A or 8BA.

Years ago, reconditioned connecting rods could actually be purchased outright. The TRW CR-1240 and the Toledo Steel CR-310 connecting rods have a 91A casting number for use with the small-journal (1.998 to 1.999 inches) crankshaft in a 1939 Ford flathead V-8.

We installed a 1949–1953 Ford 6303 crankshaft in the Model 59 engine. The best connecting rods to use with this crankshaft are the 1949–1953 Ford 6200 forged-steel I-beam model (casting number 8BA6205A or 8BA). These connecting rods are 7.00 inches long (center to center), have an average weight of 532 grams, and are designed for use with nonfloating connecting rod bearings. All the Ford flathead V-8 connecting rods were fitted with 3/8-inch N.F. connecting rod nuts.

The 1949–1953 Ford and Mercury crankshafts have two holes in each connecting rod journal for the lubrication of the nonfloating connecting rod bearings. The 1946–1948 Ford and Mercury crankshafts have one hole in each connecting rod journal for the lubrication of the full-floating connecting rod bearings. The only exceptions to this are the 1946–1948 Canadian Ford and Mercury crankshafts, which have two holes for the lubrication of the full-floating connecting rod bearings. Nonfloating connecting rods and bearings cannot be used with a full-floating

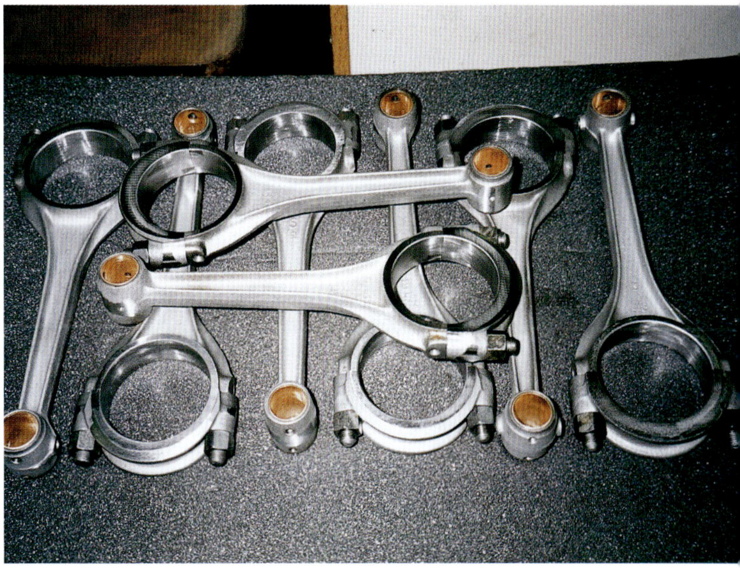

This set of 1949–1953 Ford 6200 Model 8BA6205A connecting rods was installed in the Model 59 engine. They have been shot peened, aligned, Magnafluxed, resized, rebushed, pin fitted, and balanced.

crankshaft that has one hole in each connecting rod journal. Large-journal (2.138 to 2.139 inches), full-floating connecting rods and bearings can be used with a non-floating crankshaft that has two holes in each connecting rod journal.

Magnaflux and Shot Peen

The connecting rods should be shot peened to remove stress. Do not shot peen the connecting rod studs; mask them with duct tape. Then have the connecting rods Magnafluxed to ensure there are no hidden flaws. Purchase 16 new Pioneer CRN-731 connecting rod nuts (about $3.25). The estimated cost to Magnaflux eight connecting rods is $30.

Align, Resize, and Rebush

The connecting rods should be aligned using a Sunnen Model TN-111 quick-check rod aligner tool, or similar instrument, and a connecting rod vise. Aligning ensures that the connecting rods are vertically straight in both the front and side position, and not twisted or bent.

The crankshaft end (big end) of the connecting rods should be resized on a Sunnen heavy-duty precision-honing machine (Powerstroker Model LLB-1810) or similar machine. This procedure will ensure the crankshaft end (big end) of the connecting rods is perfectly round and all the connecting rods are the same length (center to center). The estimated cost to resize eight connecting rods is $80.

After 50 years of faithful service, it was time to install new wrist pin bushings in the connecting rods. Sealed Power (Federal Mogul) manufactures two wrist bushings for the Ford flathead V-8 connecting rods. The Sealed Power 9024VA wrist pin bushing is designed for a wrist pin bore of 0.8125 inch. The Sealed Power 9024XBS wrist pin bushing is designed for a wrist pin bore of 0.819 inch. If the Sealed Power 9024XBS wrist pin bushings will be used, the wrist pin end (small end) of the connecting rods must be honed in order for the bushings to be installed. (This is not to be confused with the actual honing of the wrist pin bushing after it is installed.)

The pistons should be delivered to the machine shop with the connecting rods in order for the wrist pins to be properly fitted to the connecting rods. The *Ford Motor Company*

The crankshaft end (big end) is being honed on a Sunnen heavy-duty precision-honing machine (Powerstroker Model LLB-1810).

The Sunnen Model TN-111 connecting rod alignment tool ensures that the 1949–1953 Ford 6200 Model 8BA connecting rod is not bent or twisted.

This Sunnen heavy-duty precision-honing machine (Model LLB-1499) is honing the wrist pin end (small end) of the Ford flathead V-8 connecting rod.

of Canada Master Repair Manual specifies a connecting rod wrist pin clearance of 0.001 to 0.003 inch.

Hone the wrist pin end (small end) of the connecting rod on a Sunnen heavy-duty precision-honing machine (Model LLB-1499) or similar machine. The estimated cost for rebushing and wrist pin–fitting eight connecting rods is $120 (not including the cost of the wrist pin bushings).

De-Beam

It is not recommended to de-beam Ford flathead V-8 connecting rods for street-performance use. There is not a lot of material to play with on these connecting rods, and some things are best left alone. If the connecting rods were to be boxed, de-beaming might be considered. It is cheaper today to purchase a set of SCAT Enterprises 2-239-7000-2000 H-beam connecting rods than it is to box a set of flathead connecting rods, and the H-beam connecting rods are much stronger than boxed flathead connecting rods.

Connecting Rod Bearings

Clevite 77 (Dana Corporation) does not market full-floating connecting rod bearings for the 1946–1948 Ford flathead V-8 Model 59 engine. Federal Mogul currently offers full-floating connecting rod bearings in standard size, 0.010-inch oversize, 0.020-inch oversize, 0.030-inch oversize, and 0.040-inch oversize for the 1946–1948 Ford flathead V-8 Model 59 engines. It might be possible to obtain 1946–1948 Model 59 full-floating connecting rod bearings in other sizes from Jitney Auto Parts or Egge Parts House. The spread (end width) for the 1946–1948 Ford flathead V-8 large-journal, full-floating connecting rod bearings is 2.355 inches. Because full-floating connecting rod bearings will not be used in any of the engines described in this book, I will not provide a lengthy explanation of the spread (end width). (An explanation can, however, be found in my first flathead book.)

As previously mentioned, our Model 59 engine was fitted with 1949–1953 Ford flathead V-8 connecting rods. The Clevite 77 CB-15P nonfloating connecting rod bearings are available in standard size, 0.010-inch oversize, 0.020-inch oversize, 0.030-inch oversize, 0.040-inch oversize, 0.050-inch oversize, and 0.060-inch oversize. The current retail price for a set of these connecting rod bearings is $160. The Federal Mogul 8-1505CAA nonfloating connecting rod bearings are available in standard size, 0.010-inch oversize, 0.020-inch oversize, 0.030-inch oversize, and 0.040-inch oversize. The current retail price for a set of these connecting rod bearings is $165.

To correctly check the connecting rod bearing clearance, insert the connecting rod bearings in the connecting rods and the connecting rod caps. Next, clamp each connecting

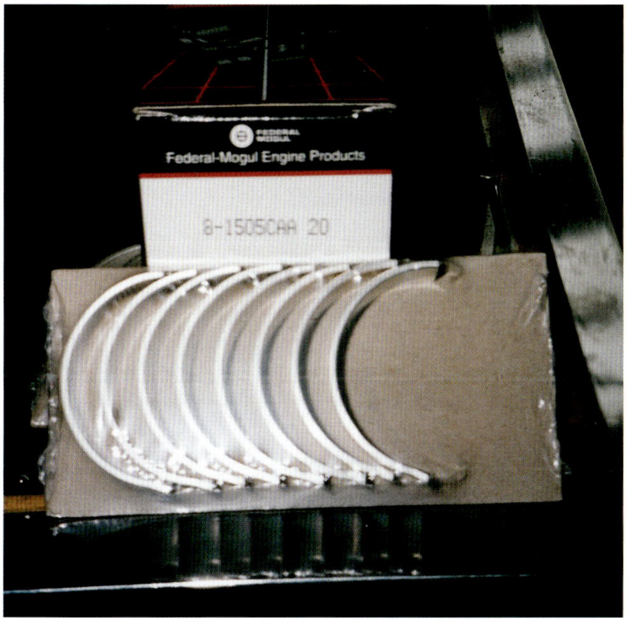

This set of Federal Mogul 8-1505CAA-20 nonfloating connecting rod bearings (0.020-inch oversize) was installed in the engine described in this section of the book.

A 1949–1953 Ford 6200 Model 8BA connecting rod is clamped in a connecting rod vise. Never clamp a connecting rod in anything other than a connecting rod vise.

The side clearance for a pair of connecting rods is being checked with a feeler gauge. Check out the meat on those Doug King Enterprises number 1 and 2 main bearing caps.

rod in a connecting rod vise and torque the connecting rod nuts to 45 ft-lb. Measure the crankshaft end (big end) bore with an inside micrometer or a dial indicator, and measure the crankshaft connecting rod journal for that connecting rod with a micrometer. The difference between these two measurements is the connecting rod bearing clearance. The *Ford Motor Company of Canada Master Repair Manual* specifies the connecting rod bearing clearance as 0.0005 to 0.003 inch. Never clamp a connecting rod into anything other than a connecting rod vise.

To correctly check the side clearance of a pair of connecting rods, clamp a pair of connecting rods in a connecting rod vise and measure the thickness of the crankshaft ends (big ends) of the two connecting rods. Use an inside micrometer to measure the distance across the crankshaft connecting rod journal for the two connecting rods. The difference between these two measurements is the connecting rod side clearance for a pair of connecting rods. This clearance should be checked again with a feeler gauge when the engine is assembled. The *Ford Motor Company of Canada Master Repair Manual* specifies the connecting rod side clearance for a pair of connecting rods as 0.006 to 0.020 inch.

Pistons

The most common pistons used in the Ford flathead V-8 engine are cast aluminum or forged aluminum. There is nothing wrong with cast-aluminum pistons; many passenger car engines equipped with cast-aluminum pistons are driven hundreds of thousands of miles without any piston problems. A forged-aluminum piston, however, is a much higher-quality piston, which is reflected in the price. Forged-aluminum pistons are the best pistons to use in a street-performance flathead engine, and some of the finest are manufactured by Ross Racing Pistons. The Ross Racing Pistons for a Ford flathead V-8 engine cost about $100 more than a set of cast pistons. Go with the quality and forget about nostalgia!

Ross Racing Pistons sells off-the-shelf stocking pistons with a 3 5/16-inch (3.3125-inch), 3 5/16-inch plus 0.030-inch (3.3425-inch), or 3 3/8-inch (3.375-inch) bore and a choice of 3 3/4-inch (3.75-inch), 4.00-inch, or 4 1/8-inch (4.125-inch) stroke. These forged-aluminum pistons come complete with heat-treated chrome-moly wrist pins, Grant cast-iron piston rings, and Spiro Lox retainers. The current retail price for a set of the forged aluminum pistons is $410.

The pistons used in the engine described in this section of the book are manufactured by Ross Racing Pistons using forged 2618 T-61 aluminum and have a slight domed top (0.187 inch). The pistons are a three-ring design and weigh 327 grams each. The top piston ring groove width is 3/32 inch. The second piston ring groove width is 3/32 inch.

This is a set of Ross Racing Pistons 80573 forged aluminum pistons with Grant Piston Rings 2360.030 cast-iron piston rings. These pistons and piston rings are intended for a 3 5/16-inch plus 0.030-inch (3.3425-inch) bore and 3 3/4-inch (3.75-inch) stroke.

The oil ring groove width is 5/32 inch. The chrome-moly steel straight-wall wrist pins weigh 120 grams each. Spiro Lox retainers are supplied with the pistons.

These off-the-shelf stocking pistons are designed for a 3 5/16-inch plus 0.030-inch (3.3425-inch) bore and a 3 3/4-inch (3.75-inch) stroke. The Grant Piston Rings 2360.030 cast-iron top and second compression ring piston ring set were used. These piston rings are not considered file fit rings and usually do not require additional gapping. On most occasions, they can be installed as they come out of the box. Grant Piston Rings recommends a piston ring gap for its cast-iron piston rings of 0.003 to 0.004 inch times the cylinder bore for the top and second compression rings. The piston ring gap for the engine described here would be 0.012 inch top ring, 0.012 inch second compression ring, and 0.014 inch for the oil ring.

To check the end gap of the pistons rings, place a piston ring approximately 1 inch from the top of the cylinder in which it will be used. The piston ring must be perfectly square in the cylinder in order to obtain an accurate measurement. Use a feeler gauge to determine the end gap of the piston ring. If the end gap is in accordance with the end gap specified by the piston ring manufacturer, move on to the next piston ring and continue until you have checked all the piston rings for all the cylinders.

If the piston ring end gap is not wide enough, the ends of the piston ring will have to be filed until the correct end gap is obtained. File the piston rings using a rotary piston ring filing tool in order to ensure the ends of the piston rings remain parallel to each other.

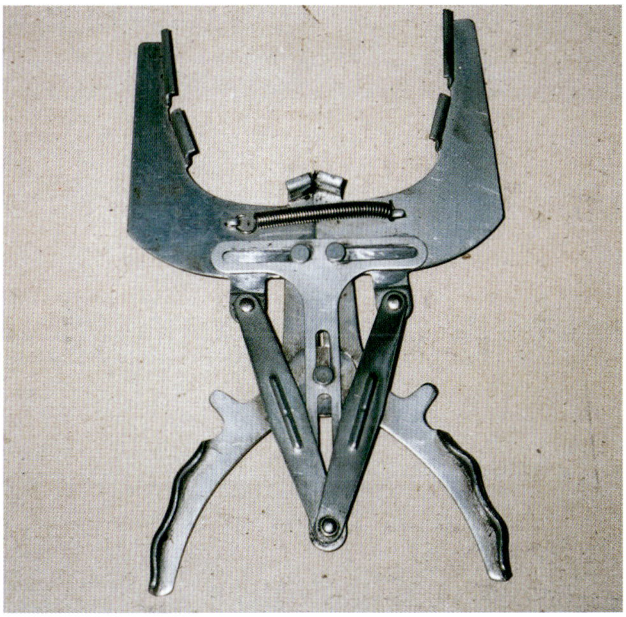

Use a piston ring installation tool to avoid gouging the pistons or breaking the piston rings when they are being installed on the pistons.

After all the piston rings are correctly gapped, install them on the pistons using a piston ring installation tool. This tool expands the piston rings enough to install the piston rings on the piston without scratching and gouging the pistons or breaking the piston rings. Prior to installing the piston and connecting rod assemblies in the engine, turn the piston rings individually on the pistons within the manufacturer's recommended arc. Turn and position each of the three piston rings on the piston so the end gap is not in the same location for all three piston rings. This is also known as preferred ring gap location. Install the piston and connecting rod assemblies in the engine block using a piston ring compressor. This tool compresses the piston rings evenly around the piston. Then insert the piston and connecting rod assembly in the cylinder by gently tapping the piston top with the handle end of a rubber mallet or a piece of wood. Use plastic connecting rod bolt protectors when installing the piston and connecting rod assemblies so as not to nick the crankshaft connecting rod journals.

Use a piston ring compressor to evenly compress the piston rings around the piston when installing the piston and connecting rod assemblies.

Balancing

One of the most important steps in building a reliable street-performance engine is balancing the rotating assembly. I have seen balancing jobs performed by backyard mechanics that removed large chunks of material from the wrist pin bosses and connecting rods—it looked like a hoard of rats had been chewing on them. Proper engine balancing is an art that should be left to the professionals.

Weigh the pistons on a Toledo digital scale or similar instrument in order to locate the lightest piston. Lighten the other seven pistons to that weight by milling material off of the bottom of the wrist pin boss.

This Ford flathead V-8 connecting rod is being balanced on a Stewart-Warner rotating connecting rod balancer (Model 329738).

Weigh the connecting rods on a Stewart-Warner connecting rod balancer (Model 329738) or similar instrument to find the lightest connecting rod. Lighten the rest of the connecting rods to that weight by removing material from the top of the connecting rod above the wrist pin or from the sides of the connecting rods and caps with a belt sander. Material is not usually removed from the bottom of the cap of a Ford flathead V-8 connecting rod because there is a limited amount of material in that area. A connecting rod is balanced by placing the crankshaft end (big end) on the balancing machine with the wrist pin end (small end) suspended. This is called a rotating balance. The wrist pin end (small end) can be placed on the balancing machine with the crankshaft end (big end) suspended. This is called a reciprocating balance.

Balance the crankshaft using a Hines Industries digital balancer or similar machine. Weigh the pistons, wrist pins, wrist pin locks, piston rings, connecting rod bearings, connecting rods, connecting rod nuts, and an estimated amount of lubricating oil. Bolt this exact weight, in the form of an attachable bob weight, to each crankshaft journal. Then spin the crankshaft on the balancing machine. Remove additional weight by drilling material from the bottom edges of the crankshaft counterweights, or add weight by drilling and inserting heavy metal (Mallory metal) in the sides of the crankshaft counterweights. This will compensate for any imbalance.

The following is how the bob weight is calculated:

Piston weight (each): __x 1 = __ grams
Wrist pin (each): __x 1 = __ grams
Wrist pin locks (pair/piston): __x 1 = __ grams
Piston rings (set/piston): __x 1 = __ grams
Reciprocating connecting rod (each): __x 1 = __ grams
Rotating connecting rod (each): __x 2 = __ grams
Connecting rod bearings (set/rod): __x 2 = __ grams
Connecting rod nuts (pair): __x 2 = __ grams
Oil (estimated): __x 1 = __ grams
Total bob weight: __ grams

If a standard (manual) transmission will be installed, have the crankshaft pulley, flywheel, clutch disc, and pressure plate delivered to the machine shop for balancing. If an automatic transmission will be installed, have the crankshaft pulley and flexplate delivered to the machine shop for balancing. A complete V-8 engine balance will eliminate any internal vibration and ensure long bearing life, provided the oil and oil filter are changed regularly. The estimated cost for a complete V-8 engine balance is $190.

Gaskets

The Fel-Pro FS7548 complete gasket set was installed in the Model 59 engine. This item is available from most auto parts dealers and from Speedway Motors (see Appendix B: Resources), part number 344-FS7548. The Fel-Pro FS7283 complete gasket set can also be used. The cylinder head gaskets contained in the Fel-Pro FS7548 set have larger-diameter center coolant holes than the cylinder head gaskets contained in the Fel-Pro FS7283 set. Special copper-sandwiched cylinder head gaskets were installed; therefore, the size of the coolant holes in the cylinder head gaskets of these two Fel-Pro sets was not a consideration.

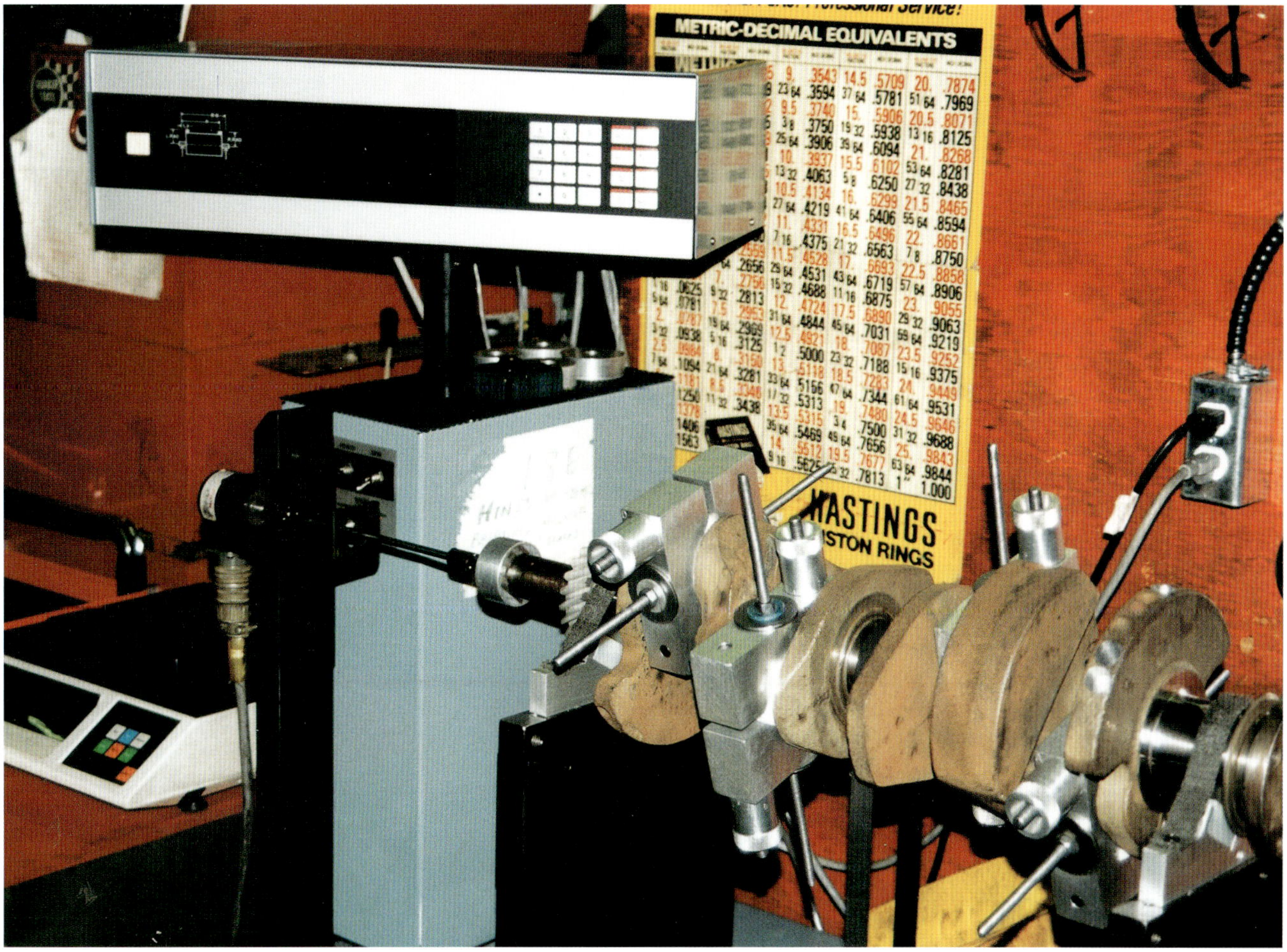

The 1949–1953 Mercury crankshaft is being balanced on a Hines Industries digital balancer. Bob weights are attached to the connecting rod journals.

Part B

- 1949–1953 Ford 6200 forged-steel connecting rods, casting number 8BA6205A; nonfloating connecting rod bearings; connecting rod length: 7.00 inches (center to center); connecting rod weight: 532 grams each; connecting rod ratio: 1.87 (with 3.75-inch stroke crankshaft); **$80**
- Connecting rods shot peened, aligned, Magnafluxed, resized, and balanced; New Sealed Power (Federal Mogul) 9024VA wrist pin bushings (0.8125-inch o.d.), installed allowing for 0.002-inch wrist pin clearance; NOS (new old stock) Canadian Ford Military LV7-FM-356004-S7 connecting rod nuts installed using Loctite and torqued to 45 ft-lb; **$286.85**
- New Federal Mogul 8-1505CAA-20 nonfloating connecting rod bearings (0.020-inch oversize), installed allowing for 0.002-inch connecting rod clearance and 0.013-inch side clearance per pair of connecting rods; **$165**
- New Ross Racing Pistons (Art Sparks original design) 80573 forged 2618 T-61 aluminum pistons; 9.1:1 compression ratio; 0.155-inch (1/8-inch plus 0.030-inch) oversize; 0.187-inch domed top; piston weight: 327 grams each; top compression ring groove width: 3/32 inch; second compression ring groove width: 3/32 inch; oil ring groove width: 5/32 inch; new Ross Racing Pistons 750-01-27 full-floating, heat-treated chrome-moly steel straight-wall wrist pins installed with new Ross Racing Pistons 4004-J Spiro-Lox retainers; wrist pin diameter: 0.750 inch; wrist pin length: 2.770 inches; wrist pin weight: 120 grams; new Grant Piston Rings 2360.030 cast-iron piston ring set installed within manufacturer's recommended arc; top compression ring gap: 0.012 inch; second compression ring gap: 0.012 inch; oil ring gap: 0.014 inch; new displacement: 263.3 ci (4.3 liters); **$409.60**
- Complete V-8 engine balance; **$190**
- New Fel-Pro FS7548 complete gasket set installed (valve guide O-rings included); **$61.95**

Connecting Rods and Pistons Total: $1,193.40

CHAPTER 4
LUBRICATION SYSTEM

Part A
Purchase

The 1946–1947 Ford flathead V-8 oil pumps were referred to as "long body" oil pumps and were equipped with spur (straight-cut) gears. These were the tallest of the flathead oil pumps and were rated at 30 psi by Ford Motor Company. The 1948 and early-1949 flathead oil pumps were of the short-body design, although they still came equipped with spur gears and were rated at 50 to 60 psi by Ford Motor Company. The later-1949 to 1953 Ford flathead V-8 oil pumps were referred to as the "short body" models; they were fitted with helical gears and were rated at 50 to 60 psi, according to the *Ford Motor Company of Canada Master Repair Manual.* The later-1949 to 1953 Ford flathead V-8 oil pumps were considered the best of all the oil pumps. However, the only oil pumps manufactured today are 1949–1953 Ford flathead V-8 replacement models, such as the Melling M-19, the Pioneer OP-19, and the Sealed Power (Federal Mogul) 224-41108. These are standard-volume oil pumps and can be purchased at most automotive parts outlets for the current retail price of $60. A Melling M-19 standard-volume oil pump was installed in the engine described in this section of the book.

This slimy creature is a pre-1948 Ford flathead V-8 long-body oil pump.

The high-volume oil pumps used in the Ford flathead V-8 engines are the Melling M-15 and the Pioneer OP-15. There is no reason to use a high-volume oil pump in a street-performance engine, when the standard-volume oil pump provides the required amount of lubrication. The current retail price for a high-volume oil pump for the Ford flathead V-8 engine is $120—an unnecessary expenditure.

This Melling M-19 standard-volume oil pump and 1949–1953 Ford truck 6615 oil pump cover and inlet tube are designed for Ford flathead V-8 engines.

Check the clearance between the gears and the top cover of the oil pump. This clearance should be 0.00025 inches (without the gasket). If there is excessive clearance, sand the top of the housing on a thick piece of glass; if the clearance is not adequate, sand the gears on a thick piece of glass. After correctly adjusting the clearance of the oil pump, thoroughly clean the oil pump, apply some motor oil to the gears, reinstall the top using a slight smear of silicone sealant on the gasket, and torque the bolts to 80 in-lb using Loctite.

The bypass spring (oil-relief valve spring) in the oil pump (or in the valve lifter gallery) controls the oil pressure. As the rpm of an engine increase, so does the volume of oil pumped through the engine, which in turn increases the oil pressure. The bypass spring limits the oil pressure. When the engine is cold, so is the oil; it is thick at that stage and difficult to pump, creating an increase in the oil pressure and causing the pressure relief valve to open, limiting the oil pressure. The *Ford Motor Company of Canada Master Repair Manual* specifies the relief valve spring tension as 12 pounds at 1.14 inches for the 1949–1953 flathead oil pump. To determine the maximum oil pressure required for your street-performance engine, allow 10 pounds of pressure for every 1,000 rpm. If the practical limit for the engine is to be 5,000 rpm, 50 psi of oil pressure will be the maximum required.

The earliest Ford flathead V-8 engine models were equipped with an oil-pressure-relief valve consisting of a ball bearing, spring, and adjustable valve nut. Around 1940, this assembly was replaced with the Ford 6663 oil relief valve plunger, the Ford 6654 oil-relief valve spring, and the Ford 6666 oil relief valve nut. These parts were located at the front of the valve lifter gallery. The Ford 6663 oil relief valve plunger had a flat spot on the head that permitted a small volume of oil to flow past it during periods of low engine rpm. The 1949–1953 Ford flathead V-8 oil pumps were fitted with a pressure-relief valve in the oil pump body.

It is not a good practice to install a 1949–1953 flathead oil pump in a 1946–1948 engine block along with the oil-relief valve already installed in the engine block. According to Ford Motor Company, the 1946–1948 flathead engine block oil-relief valve was designed to open at 30 psi. According to Melling, its 1949–1953 M-19 standard-volume oil pump pressure-relief valve is designed to open at 50 psi.

To correct the problem of two oil-pressure-relief valves in a 1946–1948 Ford flathead V-8 engine, eliminate the valve at the front of the valve lifter gallery by drilling, tapping, and installing a stainless-steel 1/8-inch NPT plug where the oil-relief valve plunger sits in the valve lifter gallery. Use Permatex aviation form-a-gasket when installing the plug. The oil-relief valve located in the 1949–1953 flathead oil pump body will then be the only oil-relief valve in the engine.

Oil Pan and Pickup

The 1946–1948 Ford 6675 oil pan with a 5-quart capacity is a very sturdy item. The rear of the oil pan forms half of the bell housing and, as a result, supports part of the weight of the transmission. The starting motor mount is incorporated into the right side (passenger's side) at the back of the oil pan. There is a road-draft opening in the right front corner (passenger's side) of the oil pan. The flywheel or flexplate should be attached to the crankshaft prior to installing the oil pan; it will not fit through the rear of the engine block when the oil pan is bolted in place. If installing a flywheel, it is easier to attach the clutch disc and the pressure plate to the flywheel before installing the oil pan. A 1946–1948 Model 59 oil pan should be fairly easy to locate and should not be expensive. (You may find one at Jitney Auto Parts.) After purchasing an oil pan, have it Redi-Stripped to ensure it is absolutely clean inside and out. Black polyurethane is the ideal color and paint to use on an oil pan, because it is extremely durable and will not easily show off road dirt. The oil pan is installed using 16 stainless-steel hex head bolts: 5/16-inch N.C. by 3/4-inch length, 16 lock washers, and 16 AN flat washers.

This 1946–1948 Ford flathead V-8 Model 59 oil pan has been hot tanked. The welding at the top near the rail is for an oil-return line fitting to be used with Ardun cylinder heads.

Remove the Ford 6751 dip stick tube boss from the oil pan prior to having the oil pan hot tanked or Redi-Stripped. Use a high-speed grinder with a carbide bit to grind off the ends of the three Ford 63359-S rivets in the oil pan, and drive the rivets out with a center punch and hammer. After the oil pan has been cleaned and before it is painted, install the dipstick tube boss with three pieces of 10-32 by 3/4-inch-long pan head screws and nuts using Loctite. A new gasket will have to be fabricated from quality gasket material and installed using silicone sealant. Grind any excess material off the ends of the three screws and stake them with a center punch.

If the Ford 6574 dip stick and 6570 dipstick tube are missing or damaged, a custom replacement can be easily fabricated. Retap the Ford 6751 dipstick tube boss with a

1/4-inch NPT tap and install a Weatherhead 202-6 brass adapter 3/8-inch tube to 1/4-inch NPT. A piece of Summit Racing Equipment 220238 stainless-steel tubing (3/8 inch o.d. by 7 5/8 inches long) is double flared on one end and installed with a Weatherhead 100-6 brass inverted flare nut 3/8-inch tube. A Trans-Dapt 4958 chrome big-block Chevrolet dip stick is reduced in length to 14 1/2 inches and installed in the tubing. Notch the full level on the dip stick with a hacksaw 2 1/4 inches from the bottom of the dip stick.

If an automatic transmission will be used with a 1946–1948 Model 59 oil pan, cut an access slot for the torque converter bolts in the bottom of the bell housing section of the oil pan. This access slot should be located 3/8 inch (0.375 inch) from the back edge of the oil pan; the slot width is 2 inches, the slot length is 6 inches, and the slot is situated in the center of the bell housing. Fabricate a sheet-metal plate to cover the slot and weld two nuts to the inside of the bell housing, permitting the slot cover to be removed.

Cut a torque converter access slot in the 1946–1948 flathead oil pan if an automatic transmission will be installed.

The Canadian Ford Military 6675 oil pan for the 1942–1948 flathead is an extremely rare and unique oil pan. Some of these oil pans have an inspection plate on the bottom (similar to some 1949–1953 Ford Truck oil pans) and some were fitted with a road-draft tube while others were not. The bell housing section is attached with bolts permitting easy access to the flywheel and clutch assembly or the flexplate and torque converter. Additionally, the flywheel or flexplate can be installed or removed with the oil pan in position. This is the ideal oil pan to use with a Model 59 engine block, although it is hard to find.

This 1942–1948 Canadian Ford military 6675 oil pan has an inspection plate in the bottom, a road-draft tube, and a detachable bell housing. This is the most desirable oil pan for a Model 59 engine block.

In order to use a Melling M-19 oil pump with the 1946–1948 flathead oil pan, a 1949–1953 Ford Truck 6615 oil pump cover and inlet tube (oil pump pickup and screen) must be installed. If you do not have one, they are available from Speedway Motors (Model 910-16259).

As soon as you purchase the gasket set, remove the cord material rear oil seals and place them in a can of motor oil. Let them soak in the oil for at least a week in order to be thoroughly saturated. These oil seals are extremely tough and dry when they are new, making installation difficult. After the oil seals have been soaked in oil, they become much more pliable. Install the oil seals by rolling them in place with a large socket.

Oil Pump Idler Gear

The 1946–1948 Ford 13-6658 oil pump idler gear cover is a cast-steel product with the Ford 6656 oil pump idler gear shaft pressed into it. This design makes it easy to remove the Ford 6655 oil pump idler gear. Removing the oil pump idler gear cover removes the oil pump idler gear as well. The oil pump idler gear is installed on the oil pump idler gear shaft with the smaller-diameter side of the oil pump idler gear facing outward from the engine block. There is a lump on the back of the oil pump idler gear cover that prevents a novice from installing the oil pump idler gear backward.

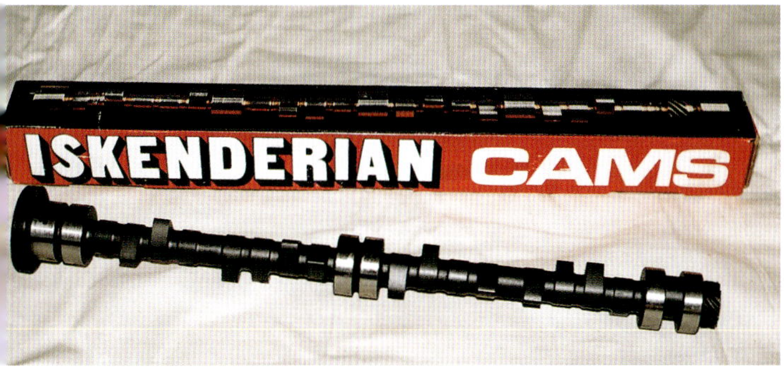

The Iskenderian 211100 Max 1 solid lifter camshaft is designed for Model 59 engines. Its advertised duration is 249 degrees intake and exhaust; duration at 0.050-inch lift: 226 intake and exhaust; lobe separation angle: 110 degrees; net valve lift: 0.350-inch intake and exhaust.

eventually sold to Ed Iskenderian (founder of Iskenderian Racing Cams).

A stock 1946–1948 Ford 6250 camshaft is not suitable for a street-performance Model 59 engine. An Iskenderian Racing Cams 211100 Max 1 solid (mechanical) lifter camshaft was installed in this Model 59 engine. This camshaft has an advertised duration of 249 degrees intake and exhaust; duration at 0.050-inch lift is 226 degrees intake and exhaust; lobe separation angle is 110 degrees; net valve lift is 0.350 inch intake and exhaust; and valve lash (hot) is 0.014 inch intake and exhaust. This is a good camshaft for a naturally aspirated smaller-cubic-inch Ford flathead V-8 motor.

The nose of the 1946–1948 Ford 6250 camshaft has a slot to accommodate the distributor drive. Some aftermarket flathead camshafts are supplied without the Ford 6254 rear camshaft gear that drives the oil pump idler gear. This gear is available at the current retail price of $10 from Speedway Motors 910-16254.

A basic understanding of camshaft terminology is helpful when trying to determine which camshaft to install in an engine. Some of the most common terms used in discussing camshafts are listed below. Catalogs from the camshaft manufacturers include a section on camshaft terminology and the mathematics associated with camshafts.

Advertised duration: Duration is the number of crankshaft degrees that the intake and exhaust valves are held open. Camshaft manufacturers specify that a camshaft has duration of so many degrees; however, they do not explain at which point of tappet (camshaft lifter) lift that duration was calculated. There is little consistency among the camshaft manufacturers in the methods used to calculate duration, making it nearly impossible to compare camshafts from different companies.

Duration at 0.050-inch lift: Several years ago, camshaft manufacturers started to measure duration at 0.050 inch of tappet (camshaft lifter) lift. This is now the standard to use when comparing camshafts.

Valve lift: This is the net lift of the valve, expressed in decimals of an inch. In a Ford flathead V-8 engine, the valve lift will be less than the camshaft lobe lift due to the valve lash, which is deducted from the camshaft lobe lift. There are no rocker arms in a flathead engine, so the valve lift is less than the camshaft lobe lift.

Overlap: Overlap is the number of crankshaft degrees that the intake and exhaust valves are open together. A camshaft with a lot of overlap creates a rough idle, less fuel economy, higher emissions, and is not recommended for a street-performance flathead engine.

Lobe separation angle: The lobe separation angle (or lobe centerline angle) is the angle between the intake and exhaust lobes at the maximum lift point, expressed in degrees. A large lobe separation angle (110–116 degrees) will result in a smooth idle and a great power range. A smaller lobe separation angle (104–110 degrees) will result in a rougher idle and a narrower power range. The engines described in this book will be driven daily on the street, which dictates that a camshaft with a larger lobe separation angle be used. The larger lobe separation angle will ensure the engine idles properly and has vacuum at idle.

Install only a new camshaft in a street-performance flathead engine. A used camshaft must have the valve lifters that were used with that camshaft matched exactly to the lobes on the camshaft, the same way they were installed in the engine. If the valve lifters are mixed up, the camshaft will be destroyed when the engine starts. Buying a used camshaft is a gamble; the camshaft might have some company's name on it, but it could have been reground or one or more of the camshaft lobes might have gone flat. Though you could take a used camshaft to a reputable camshaft company for inspection, it is generally not worth the hassle. New camshafts are not expensive, and you will know exactly what you are buying.

Camshaft Interchange

Prior to 1946, the angle between the valves in a Ford flathead V-8 engine was 101 1/2 degrees; from 1946–1953 the angle between the valves was 100 degrees. A 1946–1948 Model 59 camshaft can be used in a 1949–1953 flathead engine and vice versa. A pre-1946 flathead camshaft in a 1946–1953 flathead engine or a 1946–1953 flathead camshaft in a pre-1946 engine will not perform properly. When ordering a camshaft for a Model 59 engine from an aftermarket camshaft manufacturer, make certain that company is aware that the camshaft will be installed in a 1946–1953 flathead engine with the 100-degree angle between the valves.

The teeth on the 1946–1948 Ford flathead V-8 crankshaft and camshaft gears are slanted to provide thrust against the engine block. The teeth on the 1949–1953 Ford flathead V-8 crankshaft and camshaft gears are slanted in the opposite direction to provide thrust against the timing gear cover. If you plan to install a 1949–1953 flathead camshaft in a Model 59 engine block, you should install the 1949–1953 flathead crankshaft and camshaft gears as well. If you plan to install a Model 59 camshaft in a 1949–1953 flathead engine block, you should install the Model 59 crankshaft and camshaft gears as well. As a point of interest, some of the early 1949 Ford flathead V-8 engines were equipped with the 1946–1948 crankshaft and camshaft gears. Refer to Appendix C for information about 39 vintage camshafts.

Degree the Camshaft

There are two common practices used to degree a camshaft. Many years ago, Harvey J. Crane Jr. (the founder of Crane Cams, Inc.) started to measure duration at 0.050 inch. Known as the "duration at 0.050-inch lift method," I will explain that measurement method first.

The tools required to degree the camshaft are a 9-inch- diameter degree wheel, a 1-inch-travel dial indicator with a long stem, a magnetic adjustable dial indicator stand, and a heavy-gauge wire pointer (similar to coat hanger material). The crankshaft, connecting rod and piston assemblies, camshaft, and intake and exhaust lifters for the number 1 cylinder must be installed in order to degree the camshaft.

Locate top dead center (TDC) of the number 1 piston with the dial indicator. Install the degree wheel on the nose of the crankshaft, and attach the wire pointer to the engine block so it is as close as possible to the edge of the degree wheel. With the number 1 piston at TDC, turn the degree wheel until the pointer is at the zero-degree mark on the wheel and securely clamp the degree wheel in this position so that it cannot move. Turn the crankshaft in one direction until the piston reaches TDC; the wire pointer should be at the zero position on the degree wheel. Rotate the crankshaft in the opposite direction until the number 1 piston reaches TDC; the wire pointer should be at the zero position on the degree wheel. The degree wheel is now perfectly positioned. Remove the dial indicator from the number 1 piston.

Place the magnetic dial indicator stand on the engine block near the number 1 intake valve seat, and align the dial indicator so it passes through the valve guide bushing boss and touches the top of the number 1 intake valve lifter. The dial indicator must be at the same angle as the intake valve lifter in order to obtain an accurate reading.

Rotate the crankshaft until the maximum lift point for the number 1 intake valve lifter is located. Rotate the crankshaft one revolution past the maximum lift point—the intake valve lifter is now in the center of the base circle. Set the dial indicator to the zero position. Turn the crankshaft in the normal direction of rotation until the dial indicator reads 0.050 inch. The degree wheel will now show the number of degrees before top center (BTC) for the number 1 intake valve. Rotate the crankshaft past the maximum lift point for the number 1 intake valve lifter until the dial indicator again reaches 0.050 inch. The degree wheel now shows the number of degrees after bottom center (ABC) for the number 1 intake valve.

Remove the dial indicator and stand from near the number 1 intake valve seat and set them up near the number 1 exhaust valve seat. Align the dial indicator so it passes through the valve guide bushing boss and touches the top of the number 1 exhaust valve lifter. The dial indicator must be at the same angle as the exhaust valve lifter to obtain an accurate reading.

Rotate the crankshaft until the maximum lift point for the number 1 exhaust valve lifter is located. Rotate the crankshaft one revolution past the maximum lift point—the exhaust valve lifter is now in the center of the base circle. Set the dial indicator to the zero position. Turn the crankshaft in the normal direction of rotation until the dial indicator reads 0.050 inch. The degree wheel now shows the number of degrees before bottom center (BBC) for the number 1 exhaust valve. Rotate the crankshaft past the maximum lift point for the number 1 exhaust valve lifter until the dial indicator again reaches 0.050 inch. The degree wheel now shows the number of degrees at top center (ATC) for the number 1 exhaust valve.

All of the readings from the degree wheel for BTC, ABC, BBC, and ATC should be in accordance with the camshaft manufacturer's specification card (timing tag). Maximum engine performance will be reached if the camshaft is degreed according to those specifications.

The second accepted practice used to degree a camshaft is known as the "intake centerline method." The tools used to degree the camshaft using the "duration at 0.050-inch lift method" will be used again.

Locate TDC of the number 1 piston with the dial indicator. Install the degree wheel on the nose of the crankshaft, and attach the wire pointer to the engine block so it is as close as possible to the edge of the degree wheel.

With the number 1 piston at TDC, turn the degree wheel until the pointer is at the zero-degree mark on the wheel and securely clamp the degree wheel in this position so it cannot move. Turn the crankshaft until the number 1 piston reaches TDC; the wire pointer should be at the zero position on the degree wheel. Rotate the crankshaft in the opposite direction until the number 1 piston again reaches TDC; the wire pointer should be at the zero position on the degree wheel. The degree wheel is now perfectly positioned. Remove the dial indicator from the number 1 piston.

Place the magnetic dial indicator stand on the engine block near the number 1 intake valve seat, and align the dial indicator so it passes through the valve guide bushing boss and touches the top of the number 1 intake valve lifter. The dial indicator must be at the same angle as the intake valve lifter to obtain an accurate reading.

Rotate the crankshaft in the normal direction of rotation until the maximum lift point for the number 1 intake valve lifter is located. Set the dial indicator to the zero position. Turn the crankshaft back 0.100 inch in the opposite direction of rotation. Turn the crankshaft 0.050 inch in the normal direction of rotation. The reading on the dial indicator is now 0.050 inch before the maximum lift point of the intake valve. The degree wheel now shows the number of degrees BTC for the number 1 intake valve. Rotate the crankshaft past the maximum lift point of the number 1 intake valve until the dial indicator again reaches 0.050 inch. The degree wheel now shows the number of degrees ABC for the number 1 intake valve.

Add the number of degrees at BTC to the number of degrees at ABC and divide the total by two. The result is the intake centerline. This should be the same as the intake centerline indicated on the manufacturer's camshaft specification card (timing tag). It is extremely rare for an aftermarket Ford flathead V-8 camshaft to be degreed and the results do not match the camshaft manufacturer's specification card (timing tag). If they do not, there is likely a problem with the camshaft.

Valve Lifters

The Ford 6500 valve lifters installed in the 1946–1948 Ford flathead V-8 Model 59 engines are the same 1.00-inch diameter and 1.750-inch length as the valve lifters installed in the 1949–1953 Ford flathead V-8 engines. These valve lifters were not used in the engine described here because they are not adjustable. Instead, Iskenderian Racing Cams F85-02 Johnson solid adjustable valve lifters were installed in the Model 59 engine. These solid-body valve lifters are available directly from Iskenderian Racing Cams or from Speedway Motors 450-F8502 at a current retail price of $140 for a set.

Iskenderian Racing Cams F85-02 Johnson solid adjustable valve lifters have solid bodies and weigh 114 grams each.

The Emerson-Burkhard Manufacturing Company (Los Angeles, California) and the Witteman Company (Alhambra, California) are almost unheard of these days, but they manufactured hollow-body adjustable lifters for Ford flathead V-8 engines in the late 1940s and early 1950s. It would probably be easier to locate the *Flying Dutchman* than NOS (new old stock) sets of those lifters! Red's

These Red's Headers new Johnson hollow-body adjustable lifters weigh 83.5 grams each.

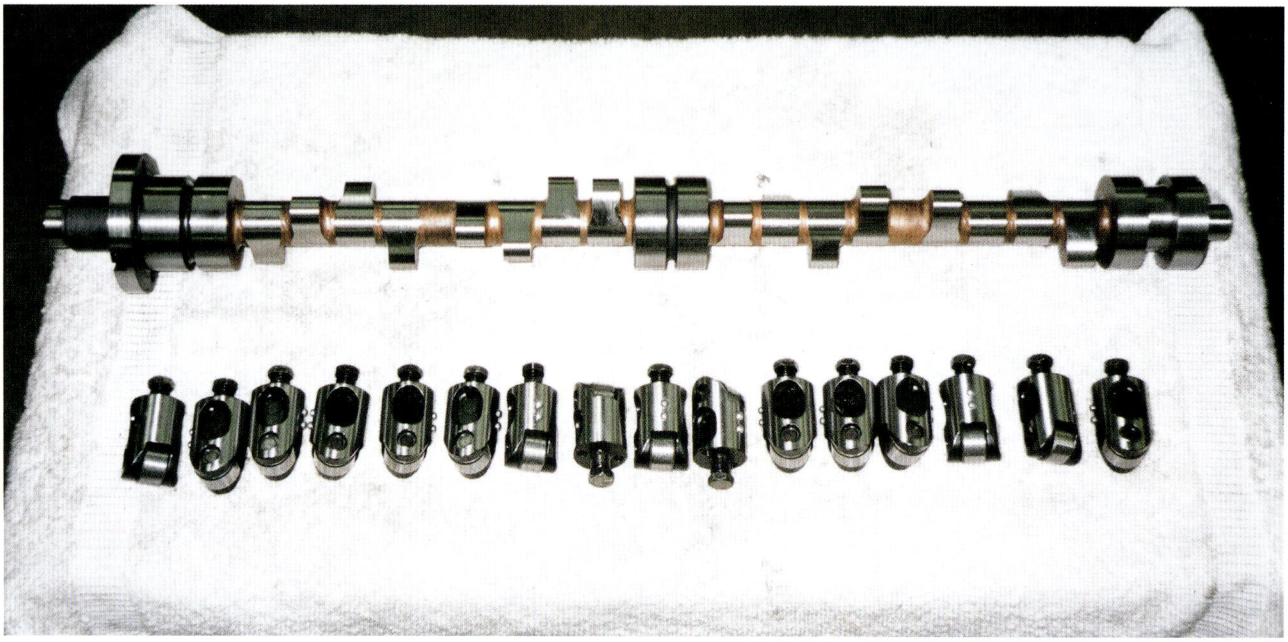

A Crower Cams & Equipment Company billet roller camshaft with the 66273 groove-lock adjustable roller lifters. The lifters are 1.00 inch in diameter, 1.540 inches long overall, and weigh 110 grams. The camshaft and lifters were installed in a Ford flathead V-8 engine to be raced on the Bonneville Salt Flats.

Headers now sells hollow-body adjustable lifters similar to the original Johnson 875 hollow-body model. These new lifters have a weight of 83.5 grams each and are available at $208 for a set of 16.

Always install new valve lifters with a new camshaft. If new valve lifters are installed with a used camshaft or used valve lifters are installed with a new camshaft, camshaft failure is almost certainly guaranteed. The Ford flathead V-8 valve lifters are referred to as "flat tappet," although the face of the valve lifter is actually convex. This permits the valve lifters to rotate, encouraging even wear on the valve lifter face and the camshaft lobe.

Valves

The Ford 6505 intake and exhaust valves for 1946–1948 Ford flathead V-8 Model 59 engines have a 1.50-inch head diameter with a 5/16-inch (0.3125-inch) stem diameter and a mushroom tip. Stainless-steel intake and exhaust valves are the best to use in a street-performance Ford flathead V-8 engine. Stainless steel is very durable and will last much longer than stock valves. One of the best sources for 1.61-inch-head-diameter by 11/32-inch-stem-diameter stainless-steel valves is Dale Wilch's RPM Catalog (see Appendix B: Resources). The 43245 stainless-steel valves are manufactured in Argentina, and their quality is as good as many of the popular domestic brands. These exhaust valves are designed for a small-block Chevrolet engine, have hardened tips, are swirl polished, and are available at the current retail price of $48 for a set of eight. A set of these stainless-steel valves was installed in the Model 59 engine described here. Exhaust valves can also be used as intake valves, but intake valves cannot be used as exhaust valves.

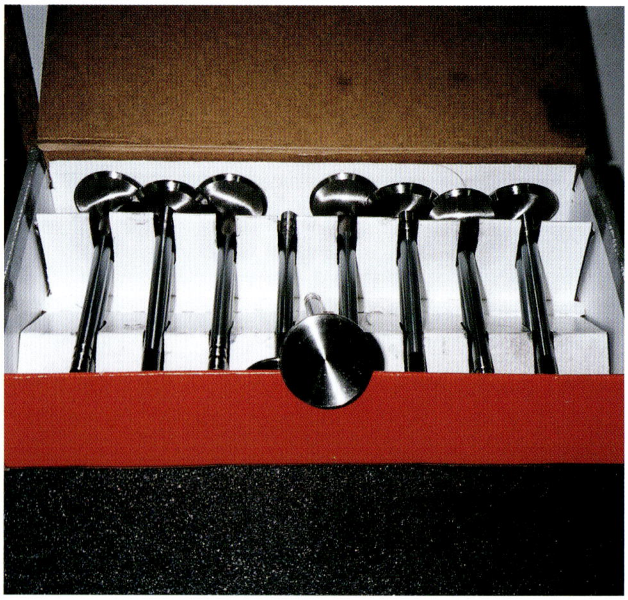

Dale Wilch's RPM Catalog 43245 stainless-steel intake and exhaust valves have a 1.61-inch head diameter and 11/32-inch stem diameter.

Grind the valves with a Kwik-Way valve grinder or similar machine to obtain a quality multiangle blueprint valve grind. A multiangle blueprint valve grind can comprise up to seven different angles, although the most common grind has three angles: 45 degrees on the valve as it sits in the valve seat insert, 30 degrees above where the valve sits, and 70 degrees below where the valve sits.

The stainless-steel valve is being ground by a Kwik-Way valve grinder.

I mentioned earlier in this chapter that the angle between the valves in the 1946–1953 Ford flathead V-8 is 100 degrees. There is also a difference in the height where the valves sit on each side of the engine block. The center of the valve head in the left side (driver's side) is 2.210 inches from the top edge of the engine block, and the top upper edge of the valve head is 0.150 inch above the deck surface in a stock engine. The center of the valve head in the right side (passenger's side) is 1.840 inches from the top edge of the engine block, and the top upper edge of the valve head is 0.050 inch above the deck surface in a stock engine. Because the valves in the left side of the engine block are 0.100 inch higher than the right side, check the valve-to-cylinder-head clearance on both sides of the engine.

Valve Guide Bushings

The Ford 6510 valve guide bushings for 1946–1948 Ford flathead V-8 Model 59 engines are a two-piece split design, which was necessary to accommodate the mushroom-tip valves. The 1949–1953 Ford 6510 one-piece valve guide bushings were installed in this Model 59 engine. If the valve guide bushings appear to be in good condition, they can be reused. The *Ford Motor Company of Canada Master Repair Manual* specifies the maximum valve-stem-to-valve-guide-bushing clearance as 0.004 inch for the intake valves and 0.005 inch for the exhaust valves. We found a good used set of these valve guide bushings that was then cleaned, glass beaded, and polished. One-piece valve guide bushings are available from Speedway Motors (part number 910-15315), or Sealed Power (Federal Mogul) VG-428 one-piece replacement valve guides are available from most automotive parts outlets.

These valve guides for the Model 59 engine were split to accommodate mushroom-tip valves.

The 1949–1953 Ford 6510 valve guide bushings can be used in the pre-1949 flathead engines, provided valves with an 11/32-inch stem diameter and 1949–1951 Ford 6514 valve spring retainers are used with them. The 1949–1953 Ford one-piece intake valve guide bushings have a groove around the middle where an oil-control O-ring was installed. The 1949–1953 Ford one-piece exhaust valve guide bushings did not have a groove around the middle; they were a solid-body design. The intake and exhaust valve guide bushings are the same diameter. The one-piece intake valve guide bushings can be used as exhaust valve guide bushings, but the one-piece exhaust valve guide bushings should not be used as intake valve guide bushings. The valve guide bushing O-rings are included in Fel-Pro's complete gasket sets. The replacement one-piece valve guide bushings available today are the intake design.

Valve Springs

Stock Ford flathead V-8 valve springs are suitable for use with a stock flathead camshaft, but they are not adequate for use with a street-performance camshaft. The Iskenderian Racing Cams 185-G single valve spring is one of the best valve springs available today for flathead engines—even other camshaft manufacturers recommend them. These top-quality valve springs are oil tempered, shot peened, and made of silicon chrome. They are available from Iskenderian Racings Cams and Speedway Motors (part number 450-185G). The current retail price for a set of these valve springs is $50.

This set of Iskenderian Racing Cams 185-G single valve springs is designed for Ford flathead V-8 engines.

The Iskenderian Racing Cams 185-G single valve springs are suitable for the majority of street-performance camshafts manufactured today. A set of these valve springs was measured at High Performance Engines on a Rinck-McIlwane precision valve spring tester. The results were:

Installed height: 85 pounds at 2.00 inches, or 120 pounds at 1.875 inches, valves closed; 208 pounds at 1.543 inches (0.332 inch lift), 218 pounds at 1.505 inches (0.370 inch lift), or 240 pounds at 1.425 inches (0.450 inch lift), valves open; 1.010 inches o.d. and 0.690 inch i.d.; overall length: 2.330 inches; coil bind: 1.345 inches; color: grey.

The Iskenderian Racing Cams 185-G single valve springs were installed in the engine described in this section of the book. The valve springs may have to be shimmed to achieve the correct installed height. Hastings VA101 valve spring shims (0.060 inch thick) or Iskenderian Racing Cams 404 valve spring shims are available for this requirement.

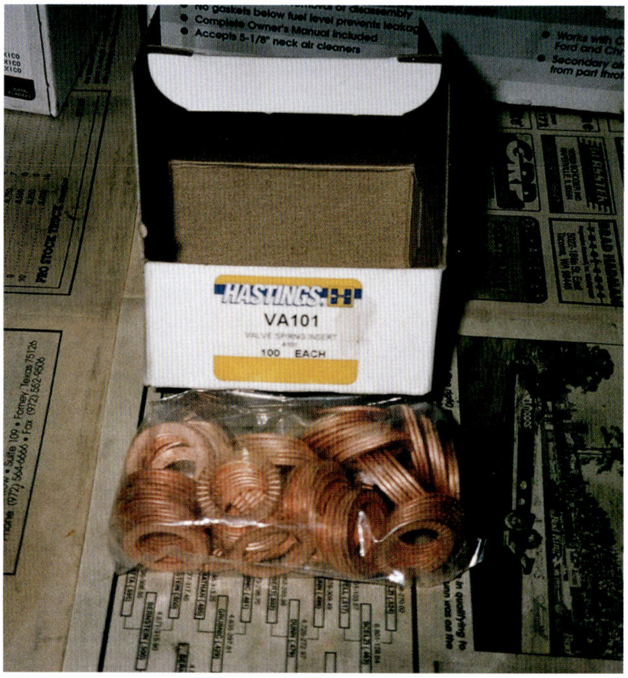

These Hastings VA101 valve spring shims (0.060 inch thick) are designed for use with Iskenderian Racing Cams 185-G single valve springs.

Valve Spring Retainers, Locks, and Seals

The 1946–1948 Ford 6514 valve spring retainers were slotted to accommodate the stock mushroom-tip valves. The 1949–1951 Ford 6514 valve spring retainers are commonly used with Iskenderian Racing Cams 185-G single valve springs. The 1952–1953 Ford 6514 valve spring retainers and 6517 sleeves are used with Iskenderian Racing Cams 4005 dual valve springs and 87-F dual valve spring retainers. On some occasions, 1952–1953 Ford 6514 valve spring retainers and 6517 sleeves have to be used with single valve springs in order to achieve the correct installed valve spring height. Ford flathead V-8 valve spring retainers are inexpensive and easy to locate.

Use a good set of valve stem locks in your street-performance Ford flathead V-8 engine. A set of heat-treated, chrome-moly, stamped-steel valve stem locks will be required when using the small-block Chevrolet stainless-steel valves with an 11/32-inch stem diameter. These are available from Crower Cams & Equipment Company (part number 86100), Manley Performance Parts (part number 13127-16), and Iskenderian Racing Cams VL-11/32-inch. The current retail price for a set of these

The Iskenderian Racing Cams 185-G single valve spring (2.330 inches long) is used in the valve assembly on the left. The Iskenderian Racing Cams 4005 dual valve spring (2.20 inches long) is used in the valve assembly on the right. Notice the two types of Ford valve spring retainers.

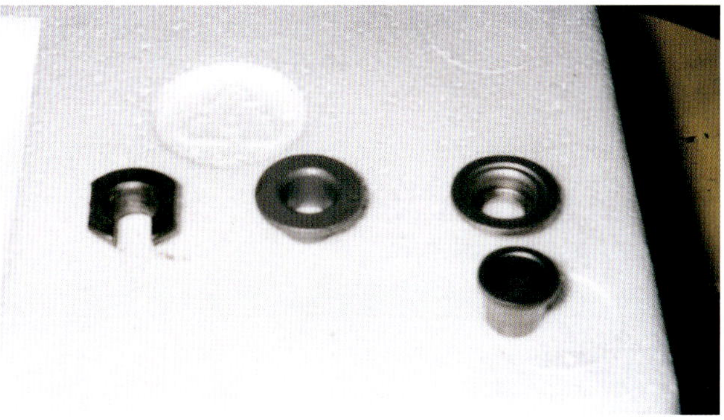

The slotted valve spring retainer on the left is a 1946–1948 Ford 6514, designed for use with mushroom-tip valves. The one in the middle is a 1949–1951 Ford 6514, designed for use with an Iskenderian Racing Cams 185-G single valve spring. The two-piece rotatable-type assembly on the right consists of a 1952–1953 Ford 6514 retainer and Ford 6517 sleeve, designed for use with Iskenderian Racing Cams 4005 dual valve springs and an 87-F dual valve spring retainer.

valve stem locks is about $10. The aftermarket stamped-steel valve stem locks are said to be stronger than stock valve stem locks. The 1949–1951 Ford 6514 valve spring retainers and Manley Performance Parts 13127-16 valve stem locks were installed in the Model 59 engine described here.

A set (16 pieces) of Ford 6512 crow's foot valve guide bushing retainers will be required. The 1946–1953 Ford flathead V-8 engines all used the same valve guide bushing retainer, so they should be very easy to find and inexpensive.

Use Manley Performance Parts 13127-16 stamped-steel 7-degree valve stem locks with small-block Chevrolet stainless-steel valves.

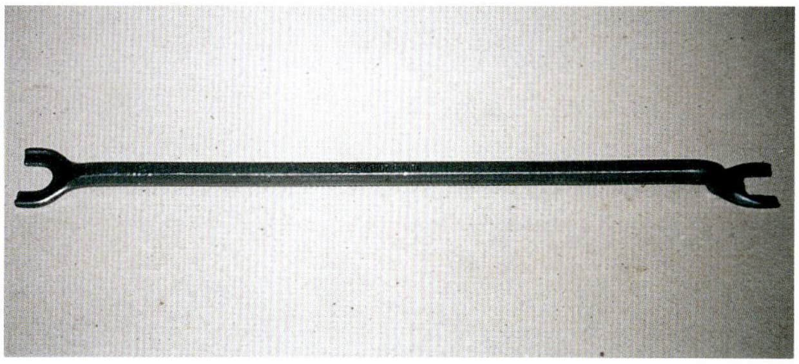

You must use a flathead valve spring compressor, like this Gray FT57 valve spring compressor, to install the valve guide bushing retainers in a flathead engine.

Camshaft Gear and Timing Cover

Do not use the 1946–1948 Ford 6256 fiber camshaft timing gear in a street-performance Model 59 engine; it will not withstand the increased stress caused by the aftermarket valvetrain. The Cloyes 2703 was an appropriate replacement fiber camshaft timing gear for the Model 59 engines, though it is no longer available from Cloyes. The 1946–1948 Ford 6256 aluminum camshaft timing gear is satisfactory for use in a street-performance Model 59 engine, as is the Melling 4706 replacement aluminum gear. This gear can be purchased from most automotive parts dealers and from Speedway Motors (part number 910-14625E). A very rare NOS 1942 Canadian Ford Military

21A-6256 cast-iron camshaft helical timing gear (with 44 teeth) was installed in the Model 59 engine described in this section of the book.

It is important to locate the Ford 6258 lock washer plate and four Ford 350400 camshaft gear bolts. These camshaft gear bolts are specially constructed to properly locate the camshaft gear on the camshaft. All the 1946–1953 lock washer plates and camshaft gear bolts are the same, so it should not be difficult to locate them at little or no cost. The four bolt holes in the flathead camshaft gear and camshaft are not symmetrical. The camshaft gear can only be installed on the camshaft in one position. The camshaft gear has a notch or dot on the face designed to align the camshaft gear with the notch or dot on the face of the crankshaft gear. These notches or dots must be aligned to correctly degree the camshaft.

"Big Ed" Winfield used to market an adjustable aluminum camshaft timing gear that permitted up to 5 degrees advance or retard of the camshaft. Red's Headers now offers an aluminum variable camshaft timing gear for pre-1949 flatheads and 1949–1953 models that permits up to 6 degrees advance or retard. These are available for \$80 or as a matched timing gear set (camshaft and crankshaft gears) with the adjustable camshaft gear for \$99.

This picture shows a new Melling 2704 crankshaft gear, a NOS 1942 Canadian Ford military 21A6019FMCC cast-iron timing gear cover, a NOS 1942 Canadian Ford military 21A-6256 cast-iron camshaft gear, and a Ford 6310 oil slinger. The back of the timing gear cover, center of the camshaft gear, and oil slinger have been painted with Glyptal G-1228A medium grey gloss enamel.

The 1946–1948 Ford flathead V-8 Model 59 engines came equipped with a cast-iron timing gear cover with two distributor bolt holes for mounting the crab-style distributor. A rare NOS 1942 Canadian Ford Military 21A6019FMCC cast-iron timing gear cover was installed on the Model 59 engine. This timing gear cover also has two distributor bolt holes. Purchase five pieces of stainless-steel hex head bolts (5/16 inch NC by 3/4 inch long) and lock washers to install the timing gear cover.

The Speedway Motors 912-S12853 one-piece front oil seal is designed for use in all 1928–1953 Ford four-cylinder and V-8 engines, but not the V-8-60.

One-Piece Oil Seal and Sleeve

The Speedway Motors 912-S12853 one-piece front oil seal is designed for 1928–1953 Ford four-cylinder and V-8 engines, with the exception of the V-8-60. This excellent front oil seal (which should have been introduced years ago) retails for \$20. There are few one-piece front oil seals currently on the market; however, some of these oil seals require an amazing amount of silicone sealant to install them. Accept no substitutes; use the Speedway Motors one-piece front oil seal. The one-piece oil seal does not have a crankshaft sleeve. Luke's Custom Machine & Design fabricates a crankshaft sleeve for use with the Speedway Motors 912-S12853 one-piece front oil seal. The Ford 6368 crankshaft sleeve has spiral grooves to lubricate the stock cord-material front oil seals. This crankshaft sleeve will not work with a one-piece front oil seal.

Luke's Custom Machine & Design crankshaft sleeve is designed for use with a Speedway Motors 912-S12853 one-piece front oil seal.

Cylinder Heads

The 1946–1948 Ford 6050 cylinder heads (casting number 59A-6050-B) were manufactured using cast iron. These cylinder heads could be used on a mild street-performance Ford flathead V-8 engine provided they were in good condition, hot tanked, Magnafluxed or pressure tested, Redi-Stripped, and surfaced. It is imperative to check the valve-to-cylinder-head clearance with any type of cylinder head, especially stock, when using oversize valves and aftermarket camshafts. Stock cylinder heads do not provide the higher compression ratio or performance look that most flathead owners like.

Champion manufactured a high-performance cast-iron cylinder head for the Model 59 engines. These cylinder heads had short fins spaced slightly over 1 inch apart. The Kogel Rocket was a very interesting cast-iron cylinder head offered for Model 59 engines. The combustion chamber design was unique in the area above the piston toward the intake valve being filled and the area above the piston downward from the exhaust valve being opened up. This cylinder head was said to provide an 8.5:1 compression ratio and had a combustion chamber volume of 63 cc. The Kogel Rocket Racer was a similar cast-iron cylinder head, but its combustion chamber volume was 58 cc, which resulted in a slightly higher compression ratio. There is enough material in the combustion chamber area of a Rocket Racer cylinder head to open it up to match a Rocket cylinder head (if a flathead owner has one of each, for example). The fins on both cylinder heads are similar in height and spacing to the Offenhauser 1068 aluminum model.

In the 1940s and 1950s, there were many aftermarket aluminum cylinder heads manufactured for the Ford flathead V-8 engine. (Frank) Baron, (Eddie) Edmunds, (Vic Sr.) Edelbrock, (Earl) Evans, (Eddie) Meyer, (Barney) Navarro, (Fred) Offenhauser, (Al) Sharp, (Bob) Tattersfield, and (Phil) Weiand—just to name a few—produced some of the best aluminum cylinder heads.

Consider purchasing used aluminum cylinder heads for a Ford flathead V-8 engine with due caution. Unless you know for certain that the set of used aluminum cylinder heads is in top condition, it is best to purchase a new set. The used cylinder heads could have been planed some time in the past, the cylinder heads might be warped, the spark plug threads could be stripped, or some of the cylinder head bolt holes could have been enlarged or drilled off center. If you do purchase a used set of aluminum cylinder heads, have them cold tanked, Magnafluxed, pressure tested, and surfaced.

The Edelbrock (see Appendix B: Resources) 1125 aluminum cylinder heads designed are for 1939–1948 Ford flathead V-8 engines. They are manufactured using 356 aluminum and T-6 tempered. The compression ratio is 8.1:1 with the stock 3 3/16 inches bore and 3 3/4 inches stroke and a block relieved to a depth of 3/16 inch. The combustion chamber volume for the Edelbrock 1125 and the 1115 (1949–1953) cylinder heads is 63 cc. These are good-looking cylinder heads with a greater spacing between the fins as compared to other finned aluminum cylinder heads. A pair of Edelbrock 1125 aluminum cylinder heads was installed on the Model 59 engine.

This picture shows the combustion chamber of the Kogel Rocket cast-iron cylinder head. Notice that the area leading up to the intake valve is filled, and the area leading down from the exhaust valve is open.

The vintage Edmunds Custom aluminum cylinder head for 1939–1948 Fords has a combustion chamber volume of 64 cc.

The Offenhauser 1068 aluminum cylinder heads are designed for 1939–1948 Ford flathead V-8 engines. These cylinder heads are available in different compression ratios based on the cylinder head numbers of: 425, 400, 375, 350, and 325; the lower the number, the higher the compression ratio (nonrelieved block). The number represents the valve-to-cylinder-head clearance, and the number of the cylinder head is supposed to be stamped near the water outlet opening, provided it hasn't been polished off. (There is a picture of an Offenhauser 1068 cylinder head on the front cover of my first flathead book.)

Red Wilson and Tony Capanna founded Wilcap Company in the late 1940s. Patrick McGuire, the new owner of Wilcap Company (see Appendix B: Resources), has started to reproduce Sharp cast-aluminum finned cylinder heads for 1939–1948 and 1949–1953 Ford flathead V-8 engines. These high-quality cylinder heads are said to have twice the water capacity of stock flathead cylinder heads. Additional material over the combustion chambers permits enough milling to accept pistons with up to 0.375-inch domed tops.

The Edelbrock 1125, Offenhauser 1068, and Sharp aluminum cylinder heads are the most economical aluminum cylinder heads available today for the Model 59 engine. The current retail price for a set of these aluminum cylinder heads is $430 for the Edelbrock model, $400 for the Offenhauser model, and $525 for the Sharp model. If you plan to install 1949–1953 Ford flathead V-8 cylinder heads on a Model 59 engine block, have the water passage hole at the very front of each cylinder head tapped and install a 1/8-inch-NPT stainless-steel plug with Permatex aviation form-a-gasket. The two water passage holes would otherwise be open and exposed.

While on the subject of later-model cylinder heads, I should mention that installing 1949–1953 Canadian Mercury 6050 aluminum cylinder heads (casting number 8CM) raised the compression ratio to 7.5:1 in the stock Ford flathead V-8 engine (239 ci). Ford Motor Company manufactured these cylinder heads for the Canadian market and other areas of North America with higher altitudes, thus they were also referred to as "Denver" heads. I recently saw a set of Model 8CM cylinder heads in fairly decent condition advertised with a selling price of $600! The pre-1949 Canadian Ford C-91A-6050-B and Canadian Ford C-3A-6050-D aluminum cylinder heads were not usually referred to as "Denver" heads; they were called "cheater" heads due to their slightly higher compression ratio and stock appearance.

A number of current engine builders insist on using cylinder head studs in a flathead engine when installing

These "cheater" heads—pre-1949 Canadian Ford C-91A-6050-B aluminum cylinder heads named for their stock appearance with 7.5:1 compression ratio—were surfaced at High Performance Engines.

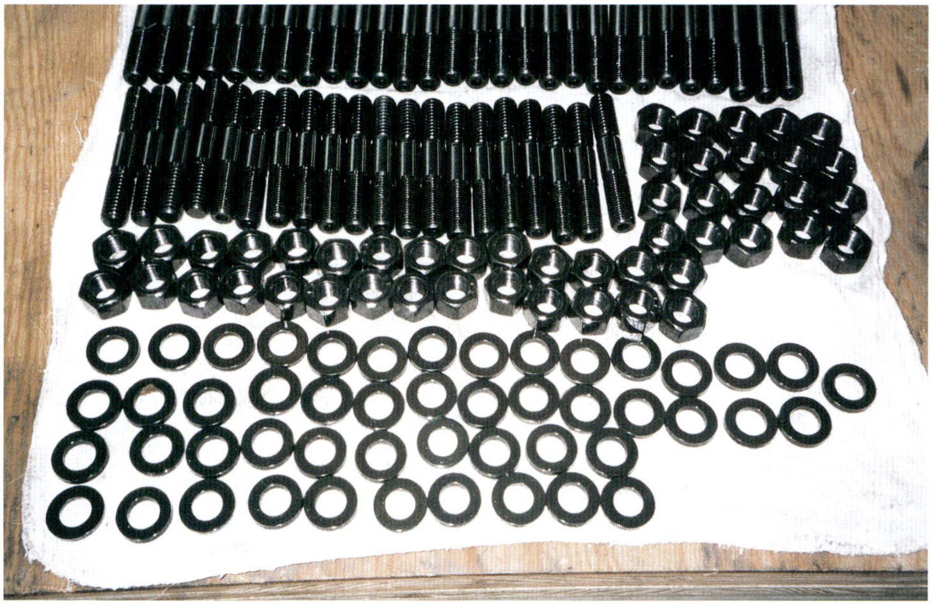

Red's Headers supplies complete sets of ARP cylinder head studs, flat washers, and nuts for the Ford flathead V-8 engines.

aluminum cylinder heads. Cylinder head studs are necessary items with a blower that produces more than 6 pounds of boost, or when the flathead engine is used for all-out racing. Aftermarket aluminum cylinder heads are torqued to 40 ft-lb, and there should not be a problem when using cylinder head bolts, provided the engine block has been parallel decked, the cylinder heads are new or have been surfaced, and the cylinder head bolts are in good condition.

Cylinder head studs are easy to install; however, the 1946–1948 Ford 6065 (short), 6066 (medium), and 6067 (long) cylinder head studs are absolute beasts to remove when they have been in the engine block for more than 50 years. If you have your heart set on using cylinder head studs, the Speedway Motors 910-15310 cylinder head stud kit (studs, washers, and nuts) is available at the current retail price of $100. ARP (Automotive Racing Products) produces some of the finest engine fasteners available today, and a complete head stud kit (studs, washers, and nuts) for the Model 59 engine is available from Red's Headers for the sum of $199. Forty-eight used 1949–1953 Ford 6066 (long) cylinder head bolts were cleaned, glass beaded, and painted for installation in the Model 59 engine block.

Use hardened-steel cylinder head bolt (or stud) washers when installing the cylinder heads on a street-performance Ford flathead V-8 engine, with or without aluminum cylinder heads. Manley Performance Parts 42102 hardened-steel flat washers are parallel ground and are ideal for use with flathead cylinder head bolts (or studs). The current retail price for a complete set (48) of the hardened-steel washers is about $20. Most flathead owners install chrome acorn nut covers for the cylinder head bolts (or studs). Speedway Motors 910-10115 chrome acorn nut covers, for a head size of 11/16 inch, are available at the current retail price of $25 for a complete set (48).

Do not have the cylinder heads on a street-performance Ford flathead V-8 engine planed. Years ago, this was a common practice used to raise the compression ratio of the engine. If the cylinder heads are planed, the upper combustion chamber area of the cylinder heads will likely have to be milled to provide adequate valve-to-cylinder-head clearance. Milling the combustion chambers will increase the volume, thereby defeating any benefit that might have been gained when the cylinder heads were planed. The engine described in this section of the book has a 9.1:1 compression ratio, which is close to the maximum compression ratio at which a street-performance engine can properly function when using today's highest-octane unleaded gasoline.

Cylinder Head Gaskets

The engine block described here was bored to 3.3425 inches (3 5/16 inches plus 0.030 inch). Fel-Pro 7548 copper-sandwiched cylinder head gaskets for 1939–1948 Ford flathead V-8 engines will fit blocks with bores up to 3.420 inches (3 3/8 inches plus 0.030 inch). Their thickness is 0.062 inch, and they have a compressed volume of 14 cc. Victor/Reinz 3036 copper-sandwiched cylinder head gaskets are also available for 1939–1948 Ford flathead V-8 engines, and they will fit blocks with bores up to 3.438 inches (3 7/16 inches). Edelbrock recommends the use of either Fel-Pro or Victor/Reinz copper cylinder head gaskets with its 1125 aluminum cylinder heads (1939–1948). Those cylinder head gaskets can be purchased from most automotive supply stores.

If copper-sandwiched cylinder head gaskets are installed in a Ford flathead V-8 engine and the engine will be stored, drain the water out of the engine block. The copper-sandwiched cylinder head gaskets will start to corrode and leak if the water is not drained from the engine block.

Water Outlets

Water outlets are cast into the top center of the 1939–1948 Ford flathead V-8 cylinder heads. If your vehicle is to be driven on the street, it should have thermostats installed. Removing the thermostats will cause the coolant in the engine to move too fast for the heat to dissipate. The correct thermostats for 1938–1948 Ford flathead V-8 engines can be purchased from Speedway Motors (part number 910-15801) at the current retail price of $16 for two.

The stamp in this pair of NOS thermostats for the Model 59 engine reads: The Fulton Sylphon Co. (Knoxville, Tenn.) 148-153.

Exhaust Baffles

I recommend installing exhaust baffles in a street-performance Ford flathead V-8 engine block. There is a common exhaust port for the two center cylinders on each side of the engine block, thus creating a situation where the exhaust gases from the two center ports collide with each other. The exhaust baffles will assist in correcting this situation. The exhaust baffle is held in place with an extended stud that replaces the stock cylinder head stud (or bolt). I also recommend drilling a small hole through the exhaust runner into the exhaust baffle and installing an Irontite plug using Loctite. This plug will ensure that the exhaust baffle does not come free or obstruct the exhaust runner if the extended stud used to secure the exhaust baffle ever loosens.

Although there are a few types of exhaust baffles on the market today, the Speedway Motors 916-15201 kit is the best one to use, as it is the original design and is available at the current retail price of $35. There is a shorter exhaust baffle available that is installed using a set screw and stud. This model requires countersinking the exhaust baffle and cutting the excess off the stud, and takes twice as long to fit in the engine block. In the early 1950s, "Honest Charley" Speed Shop (Chattanooga, Tennessee) offered a short exhaust baffle secured in place with a set screw.

Pictured are the Speedway Motors 916-15201 exhaust baffle kit for the Ford flathead V-8 engine block and the Speedway Motors 910-15315 intake valve guide bushings on a pair of stainless-steel valves.

Part B

- New Clevite 77 SH-21S camshaft bearings installed allowing for 0.002 inch camshaft clearance and 0.007 inch camshaft end play; number 1 and 2 bearings: SH-21 housing bore: 1.9275 to 1.9285 inches; number 3 bearing: SH-22 housing bore: 1.9275 to 1.9285 inches; camshaft journal diameter: 1.7965 to 1.7970 inches.; **$27.95**
- New Iskenderian Racing Cams 211100 Max 1 cast-iron alloy solid lifter camshaft with advertised duration: 249 degrees intake and exhaust; duration at 0.050-inch lift: 226 degrees intake and exhaust; lobe separation angle: 110 degrees; net valve lift: 0.350 inch intake and exhaust; valve lash (hot): 0.014 inch intake and exhaust; camshaft degreed; Ford 6524 rear camshaft gear (camshaft to oil pump idler gear) installed; **$239.90**
- New Iskenderian Racing Cams F85-02 Johnson solid adjustable valve lifters (stock 1.00-inch diameter) installed; solid body lifter weight: 114 grams, each; **$139.95**
- New Iskenderian Racing Cams 185-G aircraft-quality racing single helical coil valve springs, oil tempered, shot peened, made of silicon chrome; installed height: 85 pounds at 2.00 inches valves closed and 218 pounds at 1.505 inches (0.370-inch lift) valves open; coil bind: 1.345 inches; 1.010 inches o.d. and 0.690 inch i.d.; length overall: 2.330 inches; color: grey; 1949–1951 Ford 6514 valve spring retainers installed with new Manley Performance Parts 13127-16 valve stem locks (7-degree, heat-treated 4140 chrome-moly stamped steel, 11/32-inch stem diameter); 1949–1953 Ford 6510 one-piece valve guide bushings glass beaded, polished, and installed with 0.002-inch valve guide boss clearance and 0.002-inch valve stem clearance; 1949–1953 Ford 6512 valve guide bushing retainers installed; **$149.07**
- New (Dale Wilch's) RPM Catalog 43245 small-block Chevrolet stainless-steel 1.61-inch head diameter intake and exhaust valves, hardened tips, swirl polished, 11/32-inch stem diameter, 4.915 inches long overall; valves installed with Serdi-machined multiangle blueprint valve grind; **$96**
- NOS 1942 Canadian Ford Military 21A-6356 cast-iron camshaft helical timing gear (44 teeth) installed with Ford 6258 lock washer plate and Ford 350400 camshaft gear bolts using Loctite and torqued to 20 ft-lb; NOS 1942 Canadian Ford Military 21A6019FMCC cast-iron timing gear cover installed with new polished stainless-steel bolts and lock washers using Loctite and torqued to 15 ft-lb; timing gear cover gasket installed using silicone sealant; inside of timing gear cover painted with Glyptal G-1228A medium grey gloss enamel and outside painted with Endura EX-2C black 160 high-gloss polyurethane; **$100**
- New Speedway Motors 916-15201 exhaust baffles installed; **$34.95**
- New Victor/Reinz 3036 copper-sandwiched cylinder head gaskets installed using Permatex copper spray-a-gasket; new Edelbrock 1125 aluminum finned cylinder heads, 9.1:1 compression ratio, 63-cc combustion chamber volume, installed with 1949–1953 Ford 6066 (long) cylinder head bolts and new Manley Performance Parts 42102 heat-treated head bolt washers (7/16-inch i.d.) using Permatex aviation form-a-gasket and torqued to 40 ft-lb; new Speedway Motors 910-10115 chrome acorn nut covers (11/16-inch head size) installed; new Speedway Motors 910-15801 thermostats (180 degrees Fahrenheit) installed; valve-to-cylinder-head clearance: 0.104 inch intake and 0.132 inch exhaust (with gasket); **$755.90**

Camshaft and Cylinder Head Total: $1,543.72

CHAPTER 6
INTAKE SYSTEM

Part A
Carburetors

There are two basic types of carburetors used on street-performance Ford flathead V-8 engines. The first type is the two-barrel carburetor, the most popular models of which are the Holley 94 and Stromberg 97. The second type is the four-barrel carburetor, of which the most popular are those manufactured by Holley and Edelbrock. The engines described in this book were not equipped with two-barrel carburetors, so I shall forego any details on those types of carburetors.

The Edelbrock Performer-series nonemission four-barrel carburetors are excellent for any street engine. These silver-finish carburetors are available from 500 cfm to 800 cfm with a choice of electric or manual chokes. Prices for these carburetors range from $220 to $400. One of the true benefits of an Edelbrock Performer carburetor is that there are no gaskets below the fuel level.

The Rochester Model 4GC 486-cfm four-jet carburetor bolts directly onto an aftermarket Ford flathead V-8 four-barrel intake manifold without a carburetor adapter.

The Holley 0-8007 (Model 4160) 390-cfm four-barrel carburetor is ideally suited for a street-performance Ford flathead V-8 engine. This nonemission carburetor has vacuum secondaries and an electric choke. The current retail price for the Holley 0-8007 four-barrel carburetor is $280. If fuel economy is not a consideration and you prefer neck-snapping throttle response, purchase a Holley 0-4776C double pumper 600-cfm four-barrel carburetor. That offers far more carburetion than a flathead requires, but it is fun.

The Ford flathead V-8 aftermarket aluminum four-barrel carburetor intake manifolds were designed to accept the early Rochester (model 4GC) 486-cfm four-jet carburetor with an automatic choke. This carburetor was used on the 1955 Buick Century and Roadmaster cars, as well as many other vehicles manufactured by General Motors from 1952 through 1967. The Model 4G was the same carburetor with a manual choke. The Rochester 4G and 4GC carburetors were also produced with 553-cfm and 692-cfm ratings. Rebuilt Rochester 4G and 4GC carburetors are available today for about $275. If you plan to install a modern Edelbrock or Holley four-barrel carburetor on a flathead intake manifold, you will need a Trans-Dapt 2064 carburetor adapter, available at the current retail price of $13. Use a high-speed grinder with a carbide bit to match the four holes in the carburetor adapter with the four holes in the intake manifold.

A Rochester Model 4GC carburetor was installed on the Model 59 engine. In order to connect the automatic choke tube, a 1/8-inch-NPT hole was drilled and tapped in the top of the heat riser passage on the left-hand (driver's) side of the intake manifold. A 1/4-inch-o.d. polished stainless-steel line was fitted between the choke and the intake manifold.

PCV Valves

In cooler climates, the flathead engine is subject to serious condensation inside the engine block, particularly the valve lifter gallery. Condensation of water vapor will lead to the formation of sludge. The right front corner (passenger's side) of the oil pan on the 1946–1948 Ford flathead V-8 engines was fitted with a road-draft tube. The intake manifolds on the 1949–1953 Ford flathead V-8 engines were

The PCV valve is installed in a custom oil filler tube at the rear of a 1939–1948 Evans triple-carburetor intake manifold.

equipped with a road draft/breather tube. This open ventilation system helped somewhat, but it did not really fix the condensation problem. One of the best methods to overcome internal condensation in a flathead engine is to install a PCV (positive crankcase ventilation) valve. Canadian Ford military trucks were fitted with PCV valves from the early 1940s onward. The *Ford Truck Instruction Book (Third Edition)* supplied to defense departments of the British Commonwealth referred to the PCV valve as a "metering valve." The incorporation of the metering valve resulted in a direct-flow, sealed crankcase ventilation system. For unknown reasons, Ford Motor Company never installed this type of system on its flathead V-8 passenger vehicles.

If a Ford flathead V-8 engine is equipped with a 1939–1948 multicarb intake manifold, install a fuel pump block-off plate on the fuel pump/crankcase breather housing. Drill and tap a hole 2 inches from the front and in the top center of the intake manifold for a Weatherhead 3400-2 polished-brass 90-degree elbow (1/8-inch NPT). The Mopar Performance Parts (Chrysler Corporation) BU-1412 PCV valve (1990–1999 Plymouth Voyager with 3.0-liter Mitsubishi engine) is installed in the brass elbow (one end of the PCV valve is threaded for 1/8-inch NPT). A Weatherhead 3400-2 polished-brass 90-degree elbow (1/8-inch NPT) is installed in a vacuum source below one of the carburetor bases in the intake manifold (most aftermarket intake manifolds have vacuum fitting holes). A Weatherhead 1068-4 polished-brass male connector (1/4-inch i.d. hose to 1/8-inch NPT) is installed in the brass elbow. A 1/4-inch i.d. hose connects the PCV valve to the intake manifold fitting. Fresh air will be drawn through the fuel pump/crankcase breather housing, and the crankcase will be vented through the PCV valve. Block off the road-draft tube in the 1946–1948 flathead oil pan. The PCV valve hose can be covered with Spectre 2001A black nylabraid to give the whole installation a regal look.

If your Ford flathead V-8 engine is equipped with a 1949–1953 multicarb intake manifold, remove the fuel pump stand and install a polished aluminum block-off plate. The block-off plate is drilled and tapped in the top center to accept a Weatherhead 3400-2 polished-brass 90-degree elbow (1/8-inch NPT). A Mopar Performance Parts BU-1412 PCV valve (one end threaded for 1/8-inch NPT) is installed in the brass elbow. A Weatherhead 3400-2 polished-brass 90-degree elbow (1/8-inch NPT) is installed in a vacuum source below one of the carburetor bases in the intake manifold (most aftermarket intake manifolds have vacuum fitting holes). A Weatherhead 1068-4 polished-brass male connector (1/4-inch i.d. to 1/8-inch NPT) is installed in the intake manifold brass elbow. A 1/4-inch i.d. hose connects the PCV valve to the intake manifold fitting. Fresh air will be drawn through the oil filler/crankcase breather tube, and the crankcase will be vented through the PCV valve. Fabricate an aluminum plug to block off the road-draft/breather tube hole in the front of the 1949–1953 intake manifold and the road-draft tube in the oil pan on a 1946–1948 flathead engine.

If the flathead engine is equipped with a late-model Holley or Edelbrock four-barrel carburetor, the hose from the PCV valve connects to the PCV fitting in the base of the carburetor. The Holley 94 and Stromberg 97 carburetors were not designed with a PCV valve inlet fitting.

Intake Manifold

You will need to decide what type of intake manifold to install in the Ford flathead V-8 engine. The intake manifolds that are available are a stock type for a single two-barrel carburetor, dual carburetor, triple carburetor, four carburetors, or single four-barrel carburetor model.

The stock 1939–1948 and 1949–1953 Ford 6519 aluminum single two-barrel carburetor intake manifold is well designed. Many aftermarket multicarburetor intake manifolds have copied the runner layout of the stock flathead intake manifold. The stock intake manifold works well on a stock or slightly modified flathead engine, but it will not meet the requirements of a large-cubic-inch street-performance Ford flathead V-8 engine.

These vintage parts for a 1949–1951 Lincoln flathead V-8 are extremely rare. The aluminum intake manifold is a Nicson (Nick and son Chuck Glaviano) Engineering (Los Angeles, California) F8 dual-carburetor model. The Edelbrock finned-aluminum cylinder heads have a combustion chamber volume of 55 cc with a 9.0:1 compression ratio. "Tim the polishing guy" provided the gleam.

There were dozens of types of dual-carburetor aftermarket intake manifolds cast in the 1940s and 1950s, and they are still among the most popular models used today. Although some flathead owners swear by these intake manifolds, I have my doubts about the ability of this setup to deliver fuel economy and low emissions. The regular dual-carburetor intake manifolds permit the flathead generator to be mounted in the stock position. The super dual-carburetor intake manifolds require an offset or cylinder head mounting bracket for the generator. Edelbrock and Offenhauser currently manufacture aluminum dual-carburetor intake manifolds for the Ford flathead V-8 engines. The runner layout of the super dual-carburetor intake manifolds is superior to that of the regular dual-carburetor intake manifolds.

The triple carburetor model is one of the best intake manifolds for a street-performance flathead engine. Install progressive linkage with a three two-barrel carburetor setup for greater fuel economy, lower emissions, and excellent throttle response. Offenhauser still produces three two-barrel carburetor aluminum intake manifolds for the flathead engines, but only Stromberg 97 carburetors can be installed, due to the spacing between the carburetors. Stromberg 97 carburetors are slightly narrower than Holley 94 carburetors. The Edelbrock 1050 (Model SU-359, 1939–1948) and the Edelbrock 1050 (Model SU-349, 1949–1953) aluminum triple carburetor intake manifolds will accept Holley 94 carburetors; however, these intake manifolds are no longer produced.

Wilcap Company reproduces the Sharp "super" two-deuce and triple-carburetor aluminum intake manifolds for early (1939–1948) and late (1949–1953) Ford flathead V-8 engines. The Sharp LR3 is a triple-carburetor intake manifold, and the Sharp LR2 is the "super" two-deuce intake manifold designed for 1949–1953 engines. Holley 94 carburetors can be installed on Sharp triple-carburetor intake manifolds. These intake manifolds are superior-quality castings and are competitively priced.

The four two-barrel carburetor intake manifolds are designed strictly for racing; not even the addition of progressive linkage will make this setup suitable for street use. Edelbrock, Evans, Fenton, Sharp, Tattersfield, and Weiand used to manufacture aluminum four two-barrel carburetor aluminum intake manifolds.

The Edelbrock 452X (1939–1948) and Edelbrock 452 (1949–1953) quad-jet aluminum intake manifolds are no longer manufactured. The Edelbrock 1105 (1938–1948)

"Tim the polishing guy" tackled this Edelbrock 1050 Model SU-359 (1939–1948) intake manifold, three Holley 94 carburetors, and Offenhauser 6271 progressive linkage.

The owner bought this original Sharp LR3 triple-carburetor intake manifold for a 1949–1953 Ford 42 years ago. "Tim the polishing guy" gave the intake manifold a new lease on life.

and Edelbrock 1107 (1949–1953) are newly introduced exact-reproduction four-barrel carburetor aluminum intake manifolds. A bolt-on carburetor adapter is included with these intake manifolds to allow them to accept an Edelbrock Performer series or Holley four-barrel carburetor. Once in a blue moon, an Edmunds, Fenton QM14 (1949–1953), Navarro, or Sharp four-barrel intake manifold for Ford flathead V-8 engines magically appears at swap meets or on eBay. The Offenhauser 1079 (1932–1948) intake manifold and Offenhauser 1078 (1949–1953) four-barrel aluminum intake manifolds are still manufactured for flathead engines. Edelbrock and Offenhauser products are available from most speed-equipment mer-

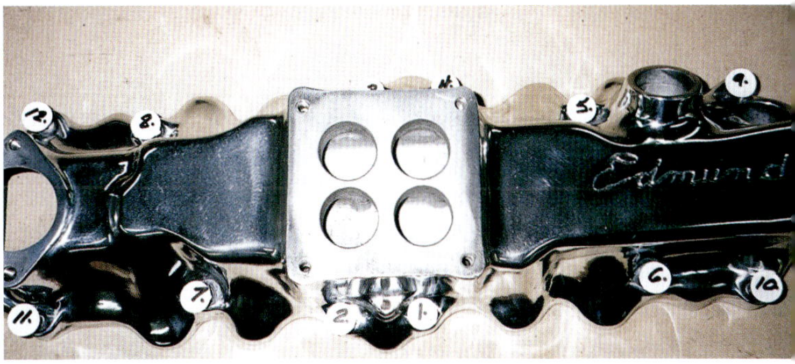

This rare 1949–1953 Edmunds four-barrel intake manifold was used on the Model 59 engine. The torque sequence recommended by Ford Motor Company is shown.

chants. The current retail price for the Edelbrock four-barrel intake manifold is $225, and the Offenhauser four-barrel intake manifold is available for $200. For this Model 59, we purchased a vintage 1949–1953 Edmunds four-barrel intake manifold in excellent condition from Jitney Auto Parts.

Fuel Lines

Any steel tubing fuel lines installed in a car must be double flared. Never install copper tubing fuel or brake lines in an automobile. Copper fatigues very easily and will fracture, leading to a disastrous situation. The best material for fuel lines is stainless-steel tubing; however, regular stainless steel cannot be double flared. Double-annealed stainless-steel tubing can be double flared and bends easily. Summit Racing Equipment is the only source for double-annealed stainless-steel tubing that I am aware of. Summit Racing Equipment SUM-220238 stainless-steel tubing is 3/8 inch diameter by 0.028 inch wall, and SUM-220256 stainless-steel tubing is 5/16 inch diameter by 0.028 inch wall. This aircraft-quality tubing is only available in 20-foot-long coils, and the current retail price is about $30. You will need a tubing cutter, tubing bender, and double-flaring kit to install steel tubing fuel lines. These tools are available from most automotive stores.

Use these tools to double flare steel tubing for fuel or brake lines. On the left is a Sears 951252 tubing cutter, in the middle is a Summit Racing Equipment 900310 double-flaring tool kit, and on the right is a Performance Tools W80675 precision tubing bender.

Fuel Pressure Regulator

The 1946–1948 Ford 9350 mechanical fuel pump for the Model 59 engine provides 1 1/2 to 3 1/2 psi fuel pressure. The fuel pump pressure can be adjusted upward by increasing the length of the fuel pump pushrod. The fuel pump pressure can be adjusted downward by reducing the length of the fuel pump pushrod. Many years ago, the Teco lifetime universal adjustable fuel pump pushrod for 1939–1948 Ford flathead V-8 engines was available. This unit replaced the Ford 48-9400A and 48-9400B fuel pump pushrods. Stock Ford flathead V-8 fuel pumps were not designed to supply fuel to multicarburetor setups, and they do not provide the required 7 psi of fuel pressure for a four-barrel carburetor. In order to have an efficient and dependable fuel system, install an electric fuel pump as close to the fuel tank as possible, in order to push the fuel and not pull it. Those flathead fanatics insisting on using a stock fuel pump will be thrilled to know that stock-style replacement fuel pumps, such as the Carter (Federal Mogul) M825, are still available for Model 59 engines from automotive parts dealers for approximately $55.

Due to their sensitivity to higher fuel pressure, operate a Stromberg 97 carburetor with a maximum fuel pressure of 2 to 2 1/2 psi. Operate a Holley 94 carburetor with a maximum fuel pressure of 2 1/2 to 3 psi. Install a fuel pressure regulator on any Ford flathead V-8 engine that is equipped with a multicarburetor setup or a four-barrel carburetor.

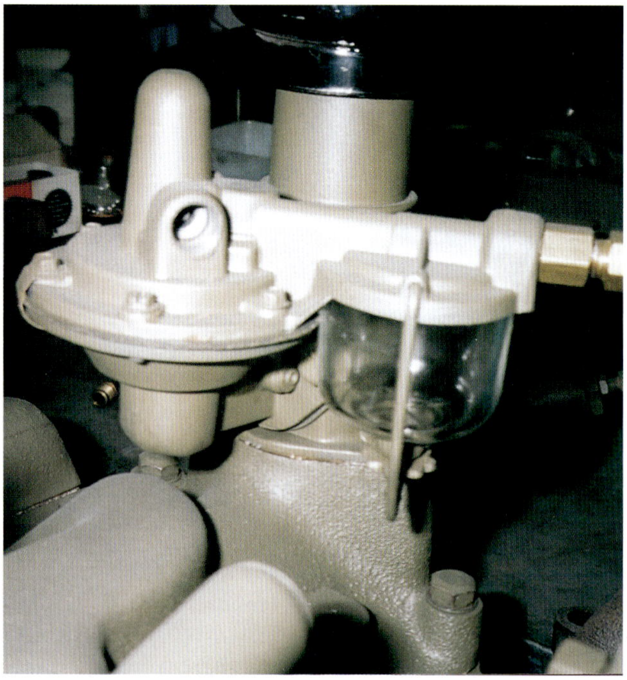

A Carter M825 replacement mechanical fuel pump for a 1946–1948 Ford Model 59 engine, painted military-olive-drab green.

There is a Holley 12-804 fuel-pressure regulator and a Holley 26-500 fuel-pressure gauge at one end of the Speedway Motors 550-6151 chrome fuel block. The three Holley 94 carburetors are mounted on an Edelbrock 1050 Model SU-359 intake manifold (1939–1948).

The Holley 12-804 fuel pressure regulator adjusts from 1 to 4 psi and is available at the current retail price of $22. Install this fuel pressure regulator when using an electric fuel pump and when the flathead engine is equipped with two or more Holley 94 or Stromberg 97 carburetors. The Holley 12-803 fuel pressure regulator adjusts from 4 1/2 to 9 psi and should be installed when using an electric fuel pump in conjunction with a four-barrel carburetor. This fuel pressure regulator is available at the current retail price of $25.

A Holley 12-803 fuel-pressure regulator with a Holley 26-500 fuel-pressure gauge is connected to an Edelbrock 1404 Performer 500-cfm four-barrel carburetor. The intake manifold is an Offenhauser 1078 (1949–1953).

Mount the fuel pressure gauge near the fuel pressure regulator to simplify fuel pressure regulator adjustment. The Holley 26-500 fuel pressure gauge has a 1 1/2-inch-diameter face and a range of 0 to 15 psi. Its current retail price is $25.

Air Cleaners

The most common types of air cleaners used today on Holley 94 or Stromberg 97 carburetors are the louvered, helmut, scoop, and velocity stack. Different air cleaners from the 1950s and 1960s occasionally appear on eBay. I have recently seen NOS Badger, Cal Custom, and Stelling & Hellings models advertised. They sold for reasonable prices, indicating that some stability may be returning to the market. There are hundreds of different types of air cleaners for the Edelbrock and Holley four-barrel carburetors, so I shall not attempt to describe or list them.

A new style of high-quality, cast-aluminum, polished air cleaner top is now available. Named the "Greenfield" air cleaner top (Ron Greenfield is the designer), they are available through Jitney Auto Parts. This model is similar to the vintage Stelling & Hellings bonnet style, except it is a slightly oval shape, which allows for greater air flow. These air cleaner tops will fit most aftermarket air cleaner bases and look terrific on Holley 94 or Stromberg 97 carburetors. New air cleaner bases are available from Speedway Motors (910-11005-Base).

Note: The air cleaner neck on the 1952–1967 Rochester Model 4G and 4GC carburetors is 4 7/32-inch diameter, not the 5 1/8-inch neck diameter found on modern Edelbrock and Holley four-barrel carburetors. The Mr. Gasket 2082 air cleaner adapter permits the installation of a 5 1/8-inch neck diameter air cleaner on these early Rochester carburetors.

This tri-power setup is equipped with Greenfield polished cast-aluminum air cleaner tops. Note the PCV valve installed at the front of the intake manifold along with three Holley 94 carburetors.

This NOS Cal Custom CC-1508-1 chrome louvered air cleaner for the 4 7/32-inch neck diameter of Rochester Models 4G and 4GC carburetors is still in its original packaging.

Bolts

The 1946–1953 Ford 20448-S intake manifold bolts were all the same. Provided the intake manifold bolts are not badly corroded, they can be reused after they have been cleaned, glass beaded, and painted. If the intake manifold bolts are missing, purchase 16 pieces (for most flathead intake manifolds) of Grade 8 or stainless steel (3/8-inches N.C. by 1 1/2-inch-long) hex head bolts and AN flat washers. In order to give a polished flathead intake manifold the finishing touch, purchase a set of Speedway Motors 910-10125 chrome acorn nut covers for the intake manifold bolts. These acorn nut covers fit a bolt-head size of 9/16 inch, and the current retail price is $15 for a set of 20 pieces.

Part B

- Edmunds polished aluminum four-barrel intake manifold (1949–1953) installed with Ford 20448-S intake manifold bolts and stainless-steel AN flat washers using Loctite and torqued to 25 ft-lb; intake manifold gasket installed using silicone sealant; new Speedway Motors 910-10125 chrome acorn nut covers for 9/16-inch head installed; Ford 9415 crankcase breather/fuel pump stand installed with new polished stainless-steel socket head bolts and lock washers using antiseize compound; fuel pump stand gasket installed using antiseize compound; new Offenhauser 5265 polished aluminum fuel pump block-off plate installed with new polished stainless-steel socket head bolts and AN flat washers using antiseize compound; fuel pump gasket installed using silicone sealant; new Speedway Motors 910-15195 crankcase breather/oil filler cap painted with PPG DGHS 9000 black high-gloss polyurethane; new polished aluminum road-draft/crankcase breather tube and crankcase breather/oil filler tube plugs installed; **$455.92**
- Rebuilt 1955 Rochester Model 4GC four-jet carburetor (486 cfm, automatic choke) installed with new Papco 264-016 engine studs (5/16 inch by 1 3/4 inches long, reduced to 1 1/2 inches) and new polished stainless-steel acorn nuts and lock washers using antiseize compound; choke tube fabricated with new Weatherhead 68-4 polished-brass male connector, 1/4-inch tube to 1/8-inch NPT; new Weatherhead 61-4 polished-brass nut, 1/4-inch tube; new Weatherhead 60-4 brass compression sleeves, 1/4-inch tube; new Summit Racing Equipment 220214 polished stainless-steel tubing, 1/4-inch o.d.; NOS Cal Custom CC-1508-1 chrome louvered air cleaner (6 1/2-inch o.d. by 1 1/2 inches high) and new K&N E-3290 air filter (6.25-inch o.d. by 1.50 inches high) installed with new polished stainless-steel acorn nut and AN flat washer using antiseize compound; **$339.97**
- New Holley 12-803 fuel pressure regulator adjustable from 4 1/2 to 9 psi installed with new Weatherhead 3325-6-2 polished-brass male hex nipple (3/8-inch NPT to 1/8-inch NPT outlet line); new Weatherhead 3220-6-4 polished-brass bushing (3/8-inch NPT to 1/4-inch NPT); new Weatherhead 1069-6 polished-brass 90-degree male elbow (3/8-inch i.d. hose to 1/4-inch NPT inlet line); new Weatherhead 3320-6-2 polished-brass bushing (3/8-inch NPT to 1/8-inch NPT); new Holley 26-500 fuel pressure gauge (1 to 15 psi, 1 1/2-inch-diameter face); all fittings installed using pipe thread sealant; **$54.35**
- New Weatherhead 3400-2 polished-brass 90-degree street elbow (1/8-inch NPT) installed with new Weatherhead 1068-4 polished-brass male connector (1/4-inch i.d. hose to 1/8-inch NPT) using pipe thread sealant; new Weatherhead 3400-2 polished-brass 90-degree street elbow (1/8-inch NPT) installed with new Motomaster 17-9259-4 PCV valve using pipe thread sealant; **$11.52**

Intake System Total: $861.76

CHAPTER 7
IGNITION SYSTEM

Part A
Distributor

The stock distributor used on the 1946–1948 Ford flathead V-8 Model 59 engine is mounted on the front of the engine with two bolts. The initial spark advance is set in these distributors by mechanical adjustment, followed by the vacuum brake adjustment. The vacuum brake adjustment is made by means of an adjustment screw while the engine is running. There was a copper vacuum line attached from the top left-hand (driver's) side of the timing gear cover to the intake manifold below the carburetor base. Due to the fact that this type of distributor uses breaker points, it will not be used on the engine described in this section of the book. If an aftermarket electronic, centrifugal (mechanical) advance distributor is used, the vacuum line hole in the timing gear cover must be plugged with a stainless-steel 3/8-inch N.F. by 1/4-inch-long socket head set screw using pipe thread sealant. The distributor is secured to the timing gear cover with two each 5/16-inch N.C. by 1-inch "L" bolts with lock washers.

The stock 1949–1953 Ford flathead V-8 distributor utilizes vacuum advance only. There is no centrifugal advance in those distributors; the vacuum advance is controlled by distributor springs and vacuum from the rear of the Holley 94 carburetor. The vacuum from the Holley 94 carburetor is a combination of the venturi vacuum and throttle plate vacuum. This system will only work with 1949–1953 flathead distributors and 1949–1953 Holley 94 carburetors (Model 8BA, Model EAB, Model 7RT, Model 8RT). If you connect the vacuum advance from a Chevrolet conversion distributor to the rear of the Holley 94 carburetor, the distributor will not function properly. The vacuum source for the Chevrolet conversion distributor must be taken from the intake manifold when using Holley 94 carburetors or from the distributor port (in the carburetor) when using the late-model Edelbrock or Holley four-barrel carburetors. The 1949–1953 Ford flathead V-8 distributors can be converted to an electronic ignition system with the Pertronix 91283 Ignitor II unit.

Every street-performance Ford flathead V-8 engine needs a good electronic ignition. The Mallory (Mr. Gasket) 5072001 magnetic breakerless distributor and Mallory 3772001 Unilite model, which is triggered by a self-contained photo-optic infrared light-emitting diode (LED) system, are excellent crab-style centrifugal (mechanical) advance distributors for a Model 59 engine. The current retail price for each of these 1942–1948 flathead distributors is $320, available through any Mr. Gasket dealer. These Mallory electronic distributors are also available for 1932–1941 Ford flathead V-8 engines.

MSD Ignitions recently introduced a new line of crab-style centrifugal (mechanical) advance, CNC-machined aluminum housing distributors equipped with a maintenance-free magnetic pickup. The MSD 8353 Pro-Billet distributor is the three-bolt model for 1939–1941 Ford flathead V-8 engines; the MSD 8351 Pro-Billet distributor is the two-bolt model for 1942–1948 Ford flathead V-8 engines. These quality distributors retail for $340 at most speed-equipment outlets. The MSD distributors must be used with MSD 6A or 6AL multiple-spark discharge ignition control or similar magnetic-triggered controls. Because Mallory and MSD distributors protrude outward approximately 4 inches from the front of the engine block, a fan mounted on the generator or generator mounting bracket will not clear them. Note: The Mallory and MSD distributors just described are all 12 volt, negative ground.

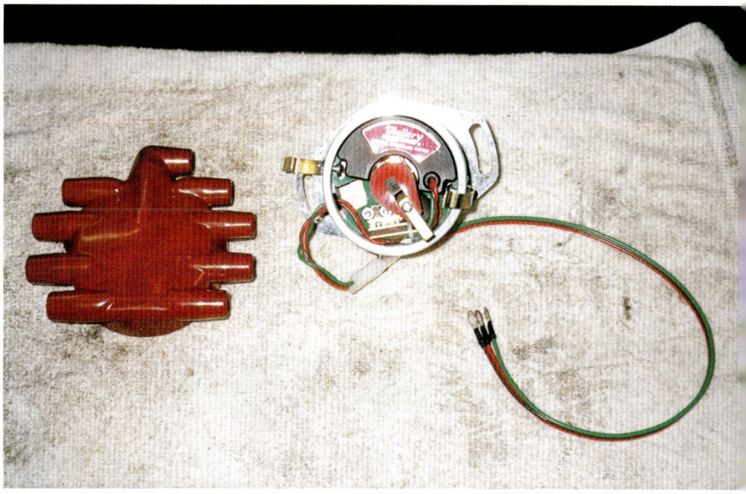

A Mallory 5072001 magnetic breakerless distributor was installed in the Model 59 engine.

The Mallory 501 Unilite conversion kit is available for factory-equipped or Mallory dual-point distributors. The Mallory 605 Unilite module and Mallory 609 magnetic breakerless module are available separately, and either of those units can be installed in a Mallory dual-point distributor. A Mallory 5072001 magnetic breakerless distributor was installed in the Model 59 engine.

Fabricate a timing pointer using a piece of stainless-steel rod bolted to the front of the timing gear cover. Set the pointer on the timing mark of the crankshaft pulley when the number 1 piston is at top dead center.

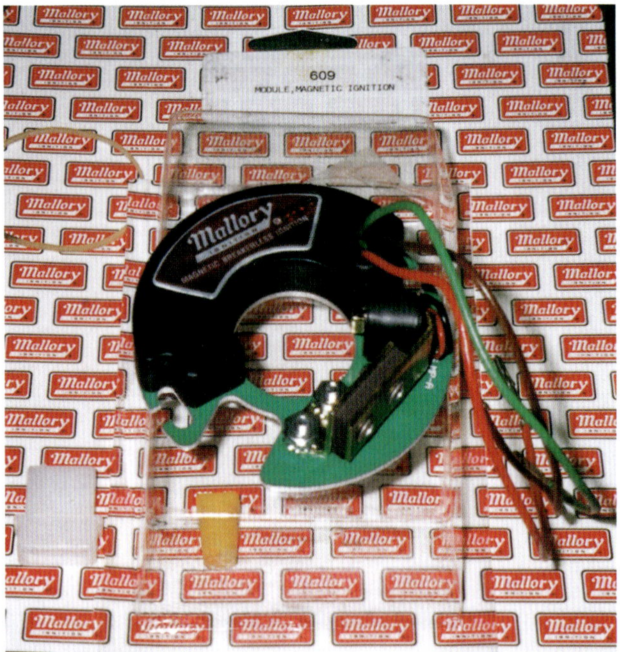

The Mallory 609 magnetic breakerless module can be used to convert any Mallory and most factory-equipped dual-point distributors.

Coil and Ignition Control

The MSD 8200 Blaster 2 chrome ignition coil with an output of 45,000 volts is ideal for a Ford flathead V-8 engine. Not all ignition coils are compatible with a multiple-spark discharge ignition control, but the MSD 8200 ignition coil is compatible. The current retail price for this ignition coil is $34. The ignition coil on 1942–1948 Ford flathead V-8 engines was attached to the Ford 21-1225 ignition coil mounting bracket at the front of the intake manifold. This bracket was designed to clear a stock location generator and will not work when an alternator is installed. A Mr. Gasket 9777 chrome ignition coil bracket can be attached to one of the intake manifold bolts. The bracket's current retail price is $12.

The Mallory 6852M Hyfire VI-A microprocessor-controlled CD ignition system is a very good unit for the Model 59 engine. The Mallory 6853M Hyfire VI-AL multispark CD ignition system with adjustable rev limiter is an excellent control box for a blown street-performance engine. The current retail price for the Mallory 6852M model is $115, and the Mallory 6853M model retails for $170.

The MSD 6A Model 6200 multiple-spark ignition control box is another very good unit for the Model 59 engine. The MSD 6AL Model 6420 multiple-spark ignition control box is equipped with rpm-limiter modules and is ideal for a blown street-performance engine. The current retail price for the MSD 6A model is $150, and the MSD 6AL retails for $196.

Spark Plug Wires and Plugs

Pertronix Performance Products 808290 Flame Thrower high-performance 8-millimeter black spark plug wires with 90-degree boots are well suited for a street-performance Ford flathead V-8 engine. These spark plug wire boots are similar in design to the original flathead spark plug wire boots. They are not the massive spark plug wire boots commonly found on OHV spark plug wires. The Pertronix Performance Products 808290 spark plug wire set is available at the current retail price of $45 from most speed shops. For a nice detailing touch, install Spectre 4245 chrome/ plastic spark plug wire separators; a set retails for $10.

The Taylor/Vertex 35071 spark plug wire is equipped with rare and obsolete Rajah AS-SOS solderless 90-degree spark plug wire terminals. The terminals are designed for use with 7-millimeter spark plug wire, although 8-millimeter spark plug wire will fit as well.

The spark plugs currently manufactured for the Ford flathead V-8 engines are the Autolite 216, Champion H-10, and NGK B-6L, all with 18-millimeter threads. A note of caution when using any type of cylinder heads on Ford flathead V-8 engines: if the upper portion of the combustion chambers in the cylinder heads has been milled for oversize valves or camshaft lift clearance, the tips of the NGK B-6L spark plugs (commonly used) may make contact with the top of the valves. If that is the case, use the NGK B-6S spark plug, which is 0.060 inch shorter in thread length than the NGK B-6L spark plug. Always coat the threads on the spark plugs with antiseize compound when they are installed in aluminum cylinder heads to prevent the threads from galling.

Alternator

GM and Chrysler alternators are most common on street-performance Ford flathead V-8 engines. The Powermaster 17127 chrome GM 65-amp, internally regulated alternator is available at the current retail price of $96. The Powermaster 17509 chrome Chrysler 75-amp, externally regulated alternator is available at the current retail price of $120. GM alternators have a smaller diameter than Chrysler alternators, although Chrysler alternators are not as long as GM alternators.

If you install a triple-carburetor intake setup on a 1946–1948 Ford flathead V-8 engine equipped with Model 59 water pumps, it will not be possible to mount a Chrysler or GM alternator on the front of the intake manifold. Because the Model 59 water pump pulleys are 1.85 inches closer to the front of the engine block than the 1949–1953 Ford Truck water pump pulleys, the Chrysler or GM alternator pulleys will not line up with the Model 59 water pump pulleys. These alternators cannot be moved any farther back on the intake manifold. Using Model 59 water pumps would require the Chrysler or GM alternator to be mounted on one of the cylinder heads. The other solution to this problem would be to fabricate custom water pump pulleys. The Speedway Motors 916-67915 bracket mounts a GM alternator on the left (driver's side) cylinder head of a flathead engine. A custom bracket would have to be fabricated to mount a Chrysler alternator on a flathead cylinder head.

The (Nippon) Denso 16241-64011 alternator is internally regulated and has an output of 35 amps. This type of alternator, available in a variety of amperages, is an extremely compact unit found on many import automobiles and Kubota tractors. It can be easily adapted to the front of any single-carburetor, dual-carburetor, triple-carburetor, or four-barrel Ford flathead V-8 intake manifold. A custom pulley can be drilled to accept the installation of a fan.

This is a Powermaster 17509 chrome Chrysler 75-amp externally regulated alternator.

This is a polished (Nippon) Denso 35-amp alternator with a polished-aluminum fan hub mounted on a polished stainless-steel bracket. Luke's Custom Machine & Design fabricated the hub and bracket. The intake manifold is a vintage Evans triple-carburetor model.

You have the option of installing a generator, which will add to the early look of a street-performance Ford flathead V-8 engine. The 1956 Ford B6A-10002-A, rated at 30 amps and 15 volts, is an excellent generator for the Model 59 engine. This generator adapts easily to the front of a flathead intake manifold by removing the pulley/fan unit, the front mounting bracket, and trimming off the rear mounting ear. A 1946–1948 Ford 21A-10139 front mounting bracket and pulley/fan unit were installed. The Model 59 pulley/fan unit screws onto the front of the 1956 Ford B6A-10002-A armature shaft without a woodruff key, and the Model 59 front mounting bracket has two bolt holes to mount a 1946–1953 Ford fan unit. The 1956 Ford B6A-10002-A generator is the same diameter as that of the 1946–1953 Ford generators. A new NTN WC87503/2A front bearing and a new Koyo 62032RSC3 rear bearing were installed. Consider chroming the generator housing, front mounting bracket, and the pulley/fan unit for that uptown look.

Starting Motor

Some flathead owners install a stock 6-volt starting motor in their street-performance engines. A 6-volt starting motor will not be harmed when used with a 12-volt electrical system, provided the starting motor is not constantly used. If it is used constantly, the 6-volt starting motor will not last.

This is a Powermaster 9507 12-volt ministarter for a Ford flathead V-8 engine.

Instead, install a Powermaster 9507 ministarter that operates on 12 volts. The compact size of this starting motor will also simplify the fabrication and installation of custom exhaust headers. Powermaster products are available at most high-performance automotive stores; the current retail price for this starting motor is $200.

Part B

- New Mallory 5072001 magnetic breakerless crab-style distributor for 1942–1948 Ford (full centrifugal (mechanical) advance; 34 degrees advance at 3,000 rpm) installed with new polished stainless-steel hex head bolts and lock washers using antiseize compound; new MSD 6200 Model 6A multiple-spark discharge ignition control and new MSD 8200 Blaster 2 chrome ignition coil (45,000 volts) installed; new Mr. Gasket 6777 chrome coil bracket installed; new polished stainless-steel timing pointer installed with new polished stainless-steel hex head bolt and lock washer using antiseize compound; **$585.23**
- New Pertronix Performance Products 808290 Flame Thrower high-performance 8-millimeter spark plug wires (black 90-degree boots) installed with new Spectre 4245 chrome professional wire separators; new NGK B-6L spark plugs installed with 0.050-inch gap and torqued to 15 ft-lb using antiseize compound; **$70**
- 1956 Ford B6A-10002-A chrome generator (rated at 30 amps and 15 volts) with 1946–1948 Canadian Ford C21A-10139 chrome front mounting bracket and Ford 10130 chrome pulley/fan unit installed with Ford 357730-S generator mounting stud, Ford 10148 washer, and Ford 356075-S stud nut using antiseize compound; new NTN WC87503/2A front bearing and new Koyo 62032RSC3 rear bearing installed; **$326.40**
- New Powermaster 9507 XS torque starting motor installed; **$189.95**

Ignition System Total: $1,171.58

CHAPTER 10
ASSEMBLY

Note: Certain procedures such as those involving bearing clearances, oil pump gear end clearance, wrist pin fitting, piston ring gaps, and installing exhaust port baffles are not mentioned in the assembly chapters of this book. These procedures were carried out prior to the final engine assembly. Also refer to Sections II and III.

Engine Block

Clean the camshaft bearings with solvent using a clean lint-free cloth. Install the camshaft bearings dry in the engine block using a camshaft bearing installation tool (this is a must), taking special care to line up the oil holes in the camshaft bearings with oil holes in the engine block. Do not lubricate the camshaft bearing bosses.

Install the oil pump idler cover gasket using silicone sealant. Lubricate the oil pump idler gear bushing and the oil pump idler gear shaft with motor oil and then install the assembly in the rear of the engine block. The smaller side of the oil pump idler gear faces outward. Install the oil pump idler gear cover bolts and lock washers using Loctite and torque them to 15 ft-lb.

Install a new stainless-steel 1/8-inch NPT pipe plug in the oil line at the face of the engine block (behind the camshaft gear) using Permatex aviation form-a-gasket. Install a new stainless-steel 1/4-inch NPT pipe plug in the oil line at the rear of the engine block (above the oil pump idler gear cover) using Permatex aviation form-a-gasket. Install a new stainless-steel 1/8-inch NPT pipe plug in place of the stock oil-pressure relief valve (located at the front of the valve lifter gallery) using Permatex aviation form-a-gasket.

Gently tap the aluminum block-off plug into the fuel pump pushrod bushing using Permatex aviation form-a-gasket. Install new polished-brass 1/4-inch NPT plugs in the water jacket drain holes in the front of each side of the engine block (behind the water pumps) using pipe thread sealant.

Install the camshaft gear on the camshaft using Loctite on the bolts and torque the bolts to 20 ft-lb. Bend the camshaft locking plate tabs over the heads of the camshaft bolts.

Lubricate the camshaft bearings and the camshaft bearing journals with Clevite 77 Bearing Guard and generously

The engine block has just been painted with PPG DGHS 9000 black high-gloss polyurethane and is now ready for assembly.

The camshaft has been installed. Note the assembly lubricant applied to the camshaft lobes.

apply the camshaft assembly lubricant (provided by the camshaft manufacturer) on the lobes of the camshaft. Gently slide the camshaft into the engine block, taking care not to nick the camshaft bearings.

Crankshaft

Install the rear main bearing oil seal retainer in the engine block using silicone sealant, and gently tap it into position with a rubber mallet. Install the rear upper cord-material oil seal in the retainer by rolling it into place with a large socket, or use the head of a large ball-peen hammer and tap the head of the hammer with another hammer. Repeat this method when installing the rear lower cord-material oil seal in the rear main bearing cap.

Clean the main bearings with solvent using a clean lint-free cloth. Line up the locking tang (tab) on the main bearings with the notch in the main bearing saddles, and install the main bearings in the engine block dry. Do not lubricate the main bearing saddles in the engine block. Use Clevite 77 Bearing Guard to lubricate the face of the main bearings after they are installed in the engine block. Line up the locking tang (tab) on the main bearings with the notch in the main bearing caps, and install the main bearings in the main bearing caps dry. Do not lubricate the main bearing caps. Use Clevite 77 Bearing Guard to lubricate the face of the main bearings after they have been installed in the main bearing caps.

The rear upper oil seal retainer is installed with the cord-material rear upper oil seal. The rear main bearing is lubricated with Clevite 77 Bearing Guard.

The crankshaft is installed, and the main bearing cap bolts are torqued to 80 ft-lb.

Install the stainless-steel connecting rod journal oil line plugs in the crankshaft using Loctite. Place the crankshaft gear woodruff key in the crankshaft using antiseize compound, and gently tap the woodruff key into position with a hammer. Apply antiseize compound to the snout of the crankshaft and slide the crankshaft gear onto the snout as far as it will go by hand. The side with the dot or line on the crankshaft gear faces outward. Line up the crankshaft gear with the crankshaft gear woodruff key, and gently tap the crankshaft gear into position using a piece of steel pipe a little larger in diameter than the crankshaft snout.

Gently lower the crankshaft into the main bearing saddles, taking special care not to nick the crankshaft bearings. Make certain to line up the timing marks on the crankshaft and camshaft gears. The crankshaft gear has a dot or a line, and the camshaft gear has a dot or a line. These marks should be vertically in line with each other. Place the main bearing caps in their correct order in the engine block with the tang (tab) in the engine block and the tang (tab) in the main bearing caps on the same side. Lubricate the main bearing cap bolts with molykote, and torque the main bearing cap bolts to 80 ft-lb. The crankshaft should now turn freely in the engine block. Slide the oil slinger onto the crankshaft snout.

Connecting Rods and Pistons

Remove the wrist pins from the pistons. Coat the wrist pin bushings in the connecting rods and wrist pins with motor oil. Some cast pistons (not Ross Racing Pistons) have an arrow on the piston head or an "F" cast into the side to indicate which side faces the front of the engine block. The cylinder numbers stamped on the connecting rods are on

Use a micrometer to measure the diameter of the connecting rod journals.

the same side as the bearing notch in the connecting rods, and these always face outward from the crankshaft. Assemble the connecting rods and pistons and install the Spiro-Lox wrist pin retainers by spreading the retainer, inserting one end into the wrist pin hole, and gently but evenly spinning the retainer into the wrist pin hole with a small screwdriver.

Install the piston rings on the pistons using a piston ring expander, and rotate each of the piston rings on the pistons so they are within the manufacturer's recommended arc, also known as the preferred ring gap location.

Clean the connecting rod bearings with solvent using a clean lint-free cloth. Install the connecting rod bearings dry in the connecting rods and the connecting rod caps. Line up the locking tang (tab) on the connecting rod bearing with the notch in the connecting rod and connecting rod cap. Apply Clevite 77 Bearing Guard to the face of the connecting rod bearings. Place plastic connecting rod bolt protectors on the connecting rod studs.

Apply motor oil to the inside of the piston ring compressor and place the piston ring compressor around the piston. Tighten the piston ring compressor and insert the connecting rod into the cylinder (with the connecting rod bearing locking tang facing outward from the crankshaft) and gently tap the piston down into the cylinder using a rubber hammer or a piece of wood. Guide the connecting studs around the crankshaft journal as you push the piston downward. After the connecting rod is on the crankshaft journal, install the connecting rod cap with the locking tang (tab) on the same side as the locking tang (tab) in the connecting rod. Install the connecting rod nuts using Loctite, and torque the connecting rod nuts to 45 ft-lb. Repeat this procedure for the rest of the connecting rod and piston assemblies.

Measure the inside diameter of the connecting rod, with the bearings installed, with a dial indicator. Note the connecting rod vise.

Install the crankshaft bolt in the crankshaft and turn the crankshaft. The rotating assembly should turn over smoothly.

Use a feeler gauge to check the side clearance for each pair of connecting rods.

Lubrication System

Install a stainless-steel 1/4-inch NPT plug inside the horizontal pressure-out oil line hole in the engine block using pipe thread sealant. Install the oil line fittings for the pressure-out oil line and drain-back oil line using pipe thread sealant.

Lubricate the oil pump drive gear with motor oil, and install the oil pump assembly in the engine block. Install the oil pump bolt and lock washer using Loctite, and torque the bolt to 15 ft-lb.

Slide the crankshaft sleeve onto the crankshaft and gently tap it into position using a piece of steel pipe and a hammer. Slide the one-piece oil seal over the crankshaft sleeve. Install the timing gear cover gasket using silicone sealant, and apply silicone sealant to the inside of the timing gear cover where the oil seal is located. Install the timing gear cover and timing gear bolts and lock washers using Loctite on the bolts, and torque them to 15 ft-lb.

A flywheel or flexplate will have to be installed in order to start the engine. The flywheel or flexplate must be installed prior to the installation of the oil pan. Install the flywheel or flexplate bolts using Loctite, and torque the bolts to 80 ft-lb.

Install the cork material gasket around the end of the rear main bearing cap using silicone sealant. Place some putty on the oil pump pickup screen and position the oil

The timing gear cover and the water pumps are installed. Note the extended studs for the exhaust port baffles.

The connecting rods are torqued to 45 ft-lb, and the oil pump assembly and flywheel are installed. Note the cork rear main bearing cap oil pan seal.

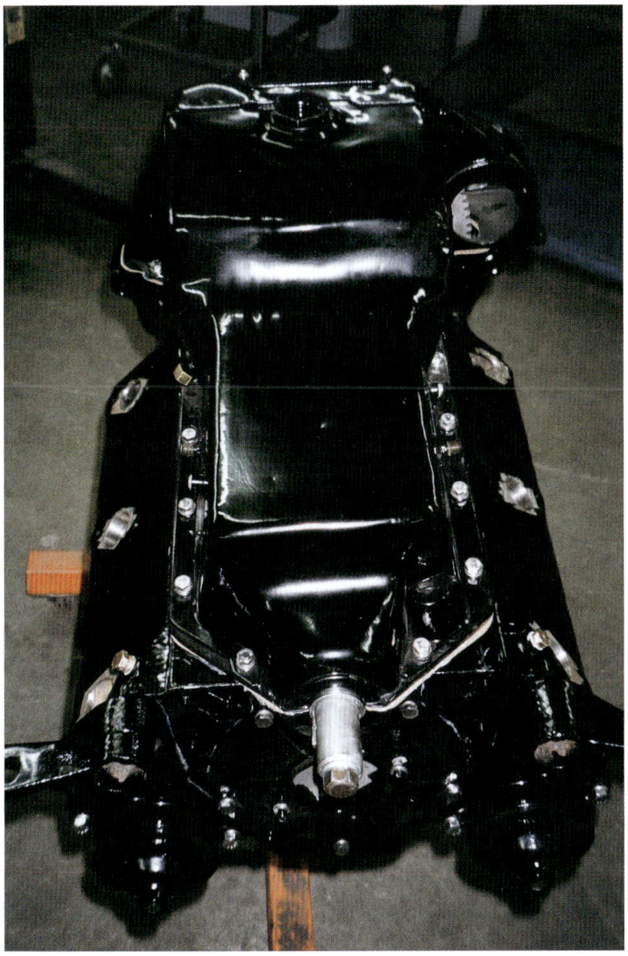

Once the Model 59 oil pan is installed, that's it for the bottom end!

pan on the engine block without the oil pan gaskets. Push down firmly on the oil pan, and then remove the oil pan. Measure the compressed thickness of the putty. The proper oil-pan-to-oil-pump-pickup-screen clearance should be 3/4 to 1 inch. If the clearance is correct, carry on. If the clearance is not correct, adjust the oil pump pickup screen arm until the correct clearance is obtained. Install the oil pan gaskets using silicone sealant. Install the oil pan and oil pan bolts, lock washers, and AN flat washers using Loctite, and torque the oil pan bolts to 15 ft-lb. Screw the dipstick tube into the oil pan using silicone sealant, and install the dip stick.

Install the oil line from the pressure-out fitting in the engine block onto the inlet side of the oil filter adapter using pipe thread sealant. Install the oil line from the drain-back line to the outlet side of the oil filter adapter using pipe thread sealant. Fill the oil filter with motor oil and install it on the oil filter adapter.

Pour the required amount of oil in the engine, and install a temporary oil-pressure gauge.

Camshaft and Cylinder Heads

Lubricate the body of the intake and exhaust valve lifter (tappet) for the number 1 cylinder with motor oil and the face of each valve lifter with the camshaft assembly lubricant supplied by the camshaft manufacturer. Install the two valve lifters, and degree the camshaft. If the results of degreeing the camshaft are in accordance with the camshaft manufacturer's specification card (timing tag), install the

Ross Racing Pistons, and 1.61-inch head diameter stainless-steel valves.

The Iskenderian Racing Cams 185-G single valve spring requires a single 0.060-inch-thick valve spring shim to achieve the recommended 2.00-inch installed height (valves closed).

rest of the valve lifters. If the results are not the same, find out why before proceeding.

Assemble the valves with the valve guide bushings, valve springs, valve spring retainers, and valve stem locks using motor oil. Install the valve guide bushing O-rings on the intake valve guide bushings. Install the valve assemblies in the engine block, and use a valve guide removal tool to compress the valve springs, allowing for the installation of the crow's foot valve guide bushing retainers. Insert a 3/16-inch drill bit (use the chuck end) or a punch into the hole drilled in the valve lifter boss, and adjust the valve according to the camshaft manufacturer's specifications. Repeat this process for all the valves. Rotate the crankshaft until the exhaust valve for a cylinder starts to open; adjust the intake valve for that cylinder. Rotate the crankshaft until the intake valve for a cylinder starts to close; adjust the exhaust valve for that cylinder. Adjust all the valves for all the cylinders in the same manner. Install

A Victor/Reinz 3036 copper-sandwiched cylinder head gasket is installed.

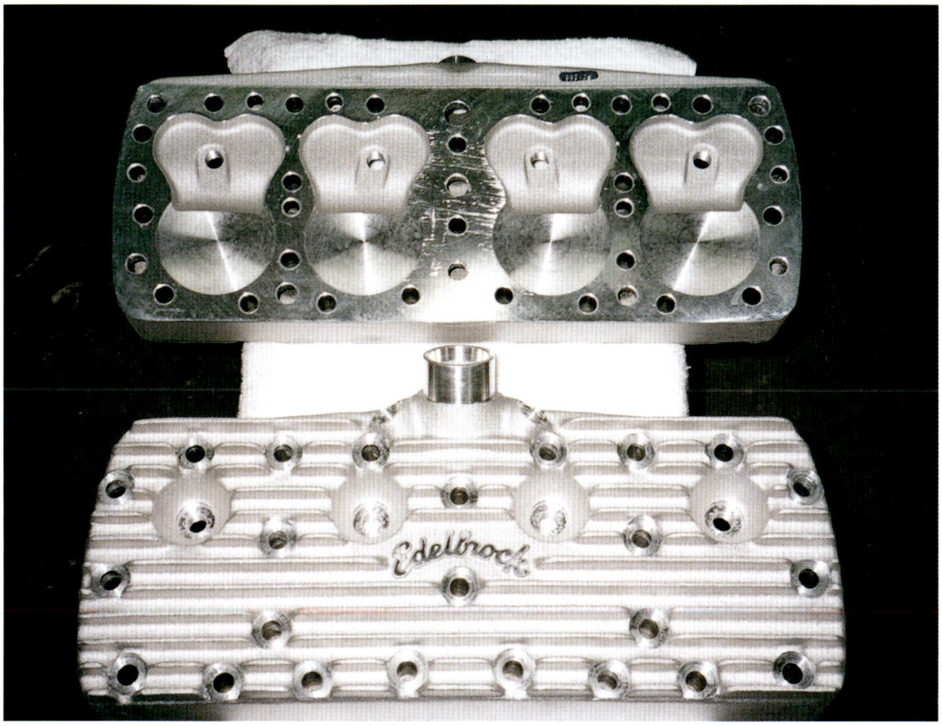

This is a brand-new set of Edelbrock 1125 aluminum cylinder heads for a 1939–1942 Ford flathead V-8 engine.

the valve lifter gallery crankcase breather tube by gently tapping it into position with a rubber hammer. Install the two valve lifter gallery oil-return hole baffles by snapping them into position. Install the two extended cylinder head studs for the exhaust port baffles using Permatex aviation form-a-gasket.

Put some putty on any intake valve and exhaust valve for the same cylinder. Place the cylinder head on the engine block, and lightly tighten the cylinder head down using two cylinder head bolts (one at each end of the cylinder head). Do not use the cylinder head gasket. Turn the crankshaft through 360 degrees at least twice. Remove the cylinder head and measure the thickness of the compressed putty. The absolute minimum valve-to-cylinder-head clearance should be 0.040 inch.

Spray the copper-sandwiched cylinder head gaskets with Permatex copper spray-a-gasket. Install the cylinder head gaskets and cylinder heads with the cylinder head

Edelbrock 1125 cylinder heads are installed, and the cylinder head bolts are torqued to 40 ft-lb.

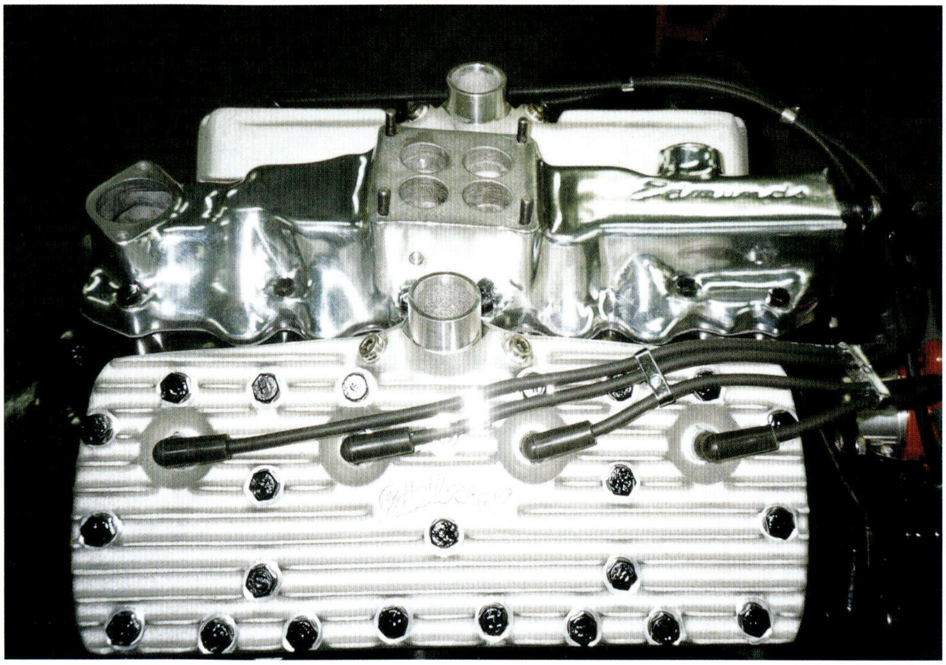

An Edmunds polished-aluminum four-barrel intake manifold is installed, and the intake manifold bolts are torqued to 25 ft-lb.

A Rochester Model 4GC four-barrel carburetor, Ford 9415 crankcase breather/fuel pump stand with the Speedway Motors 910-15195 oil filler/crankcase breather cap, and 1956 Ford B6A-10002 generator are bolted in position.

bolts and washers using Permatex aviation form-a-gasket. Torque the cylinder head bolts to 40 ft-lb (for cast-aluminum cylinder heads) using the factory-approved torque pattern (refer to Chapter 20). Do not install the chrome acorn nut covers until the engine has been initially started and the cylinder heads have been retorqued.

Install thermostats in the cylinder heads. Install the temperature-sending unit in the left (driver's) side cylinder head using pipe thread sealant. Install stainless-steel 3/8-inch NPT plugs in any unused cylinder head openings.

Intake System

Trace the intake manifold outline on the intake manifold gasket and trim the intake manifold gasket accordingly. Install the intake manifold gasket using silicone sealant. Install the intake manifold, and torque the intake manifold

The Holley 12-803 fuel-pressure regulator with Holley 26-500 fuel-pressure gauge is mounted on a stainless-steel bracket.

bolts to 25 ft-lb using Loctite. Torque the intake manifold bolts using the factory-approved torque pattern. Install the chrome acorn nut covers on the intake manifold bolts. If the intake manifold is a 1939–1948 model, install the fuel pump/crankcase breather housing using silicone sealant on the gasket, and torque the acorn nuts to 15 ft-lb using antiseize compound. If the intake manifold is a 1949–1953 model, install the fuel pump stand block-off plate using silicone sealant on the gasket, and torque the acorn nuts to 15 ft-lb using antiseize compound. Install the crankcase breather/oil filler tube and the road-draft/crankcase breather tube plugs using Permatex aviation form-a-gasket.

Install the carburetor studs in the intake manifold using antiseize compound. Install the carburetor gasket and carburetor. Install the carburetor stud nuts using antiseize compound, and tighten them snugly; do not overtighten these nuts. Install the air cleaner on the carburetor. Connect the choke tube from the carburetor to the intake manifold.

Install the fuel-pressure gauge in the fuel-pressure regulator, and mount the fuel-pressure regulator on the intake manifold. Connect the fuel line from the fuel-pressure regulator to the carburetor.

Ignition System

Install the crankshaft pulley woodruff key in the crankshaft slot using antiseize compound, and gently tap the woodruff key into position with a hammer. Apply some antiseize compound to the snout of the crankshaft and install the crankshaft pulley. The pulley should have a snug fit, but it should not require the use of a harmonic balancer installation tool. Gently tap the crankshaft pulley onto the crankshaft snout using a rubber mallet or a piece of wood and a hammer. Install the crankshaft pulley bolt and washer using Loctite, and torque the crankshaft bolt to 50 ft-lb.

Rotate the crankshaft until the engine is set with the number 1 piston at 10 degrees before top dead center (TDC), and insert the distributor in the timing gear cover with the distributor rotor at the number 1 spark plug lead position. Install the two distributor bolts using antiseize compound and tighten them snugly.

The front of the engine with the Mallory 5072001 distributor.

Gap the spark plugs to 0.050 inch, and install them using antiseize compound. Torque the spark plugs to 15 ft-lb. Lay out the spark plug wires, and cut them to length for each cylinder. The cylinders on the right side (passenger's side) are numbered 1, 2, 3, and 4 from the front of the engine block. The cylinders on the left side (driver's side) are numbered 5, 6, 7, and 8 from the front of the engine block. The firing order is: 1, 5, 4, 8, 6, 3, 7, and 2. Install the spark plug wires on the spark plugs and in the distributor cap. Install the ignition coil and ignition coil bracket. Connect the coil wire to the distributor.

Install the generator using antiseize compound on the intake manifold stud and nut. Install the starting motor using antiseize compound on the bolts.

Cooling System and Miscellaneous

Apply silicone sealant to the water pump gaskets and bolts. Install the water pumps, and torque the water pump bolts to 25 ft-lb. Install the crankshaft pulley/water pumps/generator V-belt allowing for 3/4 to 1 inch slack.

Bolt on headers and mufflers, a radiator with some water in it, wire the ignition system, and hook up a battery with a starter switch. Start the engine, and adjust the timing and the carburetor. Allow the engine to run at approximately 2,000 rpm for 30 minutes in order to break in the camshaft. After the engine is at full operating temperature, shut it off, and allow the engine to cool. Retorque the cylinder heads using the factory-approved torque pattern, and install the chrome acorn nut covers on the cylinder head bolts.

This engine has more than double the horsepower of a 1948 Ford flathead V-8 engine. There is almost a straight torque curve from 2,000 to 4,000 rpm, indicating the correct camshaft was chosen for this particular engine.

Luke Balogh fabricated these metallic-ceramic-coated equal-length exhaust headers for his 1950 Meteor (Canadian Ford).

The left side of the Model 59 engine looks as good as it runs.

SECTION II: BLOWN FRENCH FLATHEAD STREET ENGINE

CHAPTER 11
ENGINE BLOCK

A few years ago, a warehouse full of flatheads was discovered in France. The French automobile company SIMCA (Societe Industrielle de Mecanique et Carrosserie Automobile) had manufactured an improved version of the Ford flathead V-8 for the French military from 1962 to 1990, or shortly thereafter. This engine was a combination of the 1946–1948 and 1949–1953 Ford flathead V-8 engine block. A number of these flatheads and their associated engine parts have made their way to the United States.

A French copy of the 1937–1940 Ford V-8-60 engine was supposedly installed in SIMCA automobiles sold in Brazil from 1960 to 1969. Chrysler do Brasil (Chrysler of Brazil) was said to have manufactured this engine in 1969 only for the Brazilian market.

Some parties are turning to the French flathead V-8 engine block as a solution to their frustration in trying to find a used Ford flathead V-8 engine block in good serviceable condition. Keith Lee, the owner of Jitney Auto Parts, recently told me he was averaging one good Ford flathead engine block out of every seven he had located. "Perseverance" is the keyword here, and Jitney Auto Parts is an excellent source for good used Ford flathead V-8 engine blocks.

The French flathead V-8 engines are said to be used, overhauled, and ran-in tested—all without guarantee. They even come with French documentation, which should be a real treat to read! The current retail price for a complete French flathead V-8 engine with a 3 3/4-inch-stroke crankshaft (239.4 ci) is $3,000. The current retail price for a complete French flathead V-8 engine with a 4-inch-stroke crankshaft (255.4 ci) is $3,150. This price might seem appealing; however, there are many parts on these engines that will not be used to build a street-performance engine. I will explain the benefits of some of these engine parts and the negative aspects of other engine parts in this section of the book.

A local owner of a French flathead V 8 engine block approached Luke's Custom Machine & Design in June 2003 with the intention of having a blown street-performance engine built using that engine block as the foundation for the project. I decided to capitalize on a great opportunity by describing the complete buildup of this engine. Refer to Section I for a full explanation of any subheadings that are left blank in this section. The owner of this French flathead V-8 engine installed it in a daily driven 1934 Ford pickup.

Part A

I do not want to rain on anyone's parade, but I believe the French flathead V-8 engine block is a luxury and an unnecessary expenditure when good used Ford flathead V-8

This is the French flathead V-8 engine block used as an example in this book. The factory relief is clearly visible, along with the ported and polished intake ports and larger intake valve seat inserts.

engine blocks are still available. The current retail price for a new old stock (NOS) French flathead engine block is $1,800. Decent Ford flathead V-8 engine blocks can be purchased for a lot less money, and they are suitable for any type of street-performance flathead engine, blown or not. And the money saved by buying a Ford flathead engine block rather than a French flathead engine block will buy a lot of speed equipment!

The material used to manufacture French flathead V-8 engine blocks is advertised as dense, nonporous, superior steel alloy. Additionally, the engine blocks are said to be built to SAE (Society of Automotive Engineers) specifications with no casting flash or obvious core shift and to have thicker decks and cylinder walls and beefier cylinder head bolt holes than the 1946–1953 Ford flathead V-8 engine blocks. The French flathead V-8 engine block described in this section weighs exactly 230 pounds, which is 30 pounds heavier than a 1949–1953 Ford flathead V-8 engine block.

Some of these so-called improvements need to be addressed. The French flathead V-8 deck thickness averages 1/8 inch (0.125 inch) more than the deck thickness of the 1949– 1953 Ford flathead V-8. The engine block thickness in the gasket area around the timing gear cover and water pumps in the French flathead is 0.360 inch, and the thickness in the same areas of a 1949–1953 Ford flathead is 0.265 inch. The thickness of the cylinder walls in this particular French flathead with a standard bore (3 3/16 inches) is 0.255 inch; the thickness of the cylinder walls in a standard-bore (3 3/16 inches) 1946–1948 Ford flathead is 0.255 inches. This measurement appears to debunk the thicker-cylinder-wall theory in the French flathead V-8 engine block. The French flathead and Ford flathead are similar, in that some engine blocks are capable of safely handling a 3 3/8-inch (3.375-inch) bore. Very few Ford flathead V-8 engine blocks were ever safely bored to 3 7/16 inches (4.375 inches), and it seems this applies to the French flathead V-8 as well. The cylinders in the French flathead were tough to bore, indicating that the manufacturer used material with high nickel content.

This picture shows the engine's casting number (4580015) and casting date (30-06-87). The engine governor boss was originally located in the area that is now filled with liquid steel. There is still a lot of excess material on top of the bell housing.

The casting number for this French flathead V-8 engine block is 4580015, and the casting date is 30-06-87 (June 30, 1987). A casting number of 702550 has also been reported; there may well be other casting numbers. The casting number and casting date are cast into the top of the integral bell housing.

Purchase and Magnaflux

The French flathead V-8 engine block is a combination of 1946–1948 and 1949–1953 Ford flathead V-8 engine blocks. The enlarged coolant flow holes at the rear of the engine block are similar to those in 1949–1953 Ford flathead V-8 engine blocks. The integral bell housing and rear oil seal are similar to those on the 1946–1948 Ford flathead V-8 engine blocks—two of the less-desirable features that Ford Motor Company did away with in 1949.

The first thing to do when you purchase the engine block is to hot tank it. The French flathead blocks are coated with a substance similar to Cosmoline; it is impossible to properly examine an engine block when it is in that state. Even though these engine blocks are said to be NOS, they should be thoroughly examined for cracks. Have the engine block and main bearing caps Magnafluxed (technically known as "magnet particle testing"). The estimated cost to Magnaflux the engine block and main bearing caps is $50.

The engine block must have at least the rear main bearing cap attached. Because there are very few of these engine blocks in existence, there is a limited possibility of locating a replacement rear main bearing cap. New billet steel number 1 and 2 main bearing caps are available from Doug King Enterprises (see Appendix B: Resources). These main bearing caps were added to the engine described here because it was equipped with a blower.

Valve Grind

Refer to Section I

Deglaze

Refer to Section I

Cracks

Refer to Section I

Cleaning and Painting

Refer to Section I

Part B

- New old stock (NOS) 1987 French military 239.4-ci L-head V-8 engine block; casting number: 4580015; casting date: 30-06-87 (June 30, 1987); two-bolt main bearing caps; 24-stud block; stock bore: 3 3/16 inches (3.1875 inches); stock stroke: 3 3/4 inches; engine block weight (bare): 230 pounds; **$1,800**
- Engine block hot tanked and Redi-Stripped; engine block, main bearing caps, and main bearing cap bolts Magnafluxed; **$245**
- All threads retapped; main bearing caps and main bearing cap bolts shot peened; cylinder head bolt holes chamfered; valve lifter bosses deglazed allowing for 0.002-inch valve lifter clearance; valve lifter bosses drilled with 3/16-inch (0.1875-inch) hole for valve lifter adjustment; engine governor boss removed by grinding; bottom of cylinder bores ground for connecting rod clearance; new stainless-steel water jacket drain plugs (1/4-inch NPT) installed using pipe thread sealant; valve lifter gallery and front of engine block painted with Glyptal G-1228A medium grey gloss enamel; exterior surface of engine block sanded, detailed, and painted with PPG 74000 red high-gloss polyurethane; **$601.39**
- Engine block converted to full-flow oil system; **$30**
- Engine block bored 0.040-inch oversize, and cylinders honed and deglazed using Sunnen 800-series 400-grit stones; final bore: 3.2275 inches; piston-to-bore clearance: 0.006 inch, measured below bottom of wrist pin perpendicular to wrist pin; **$130**
- Engine block parallel decked to 0.010-inch average, below deck; **$120**
- Engine block align bored and align honed; new Doug King Enterprises billet steel number 1 and 2 main bearing caps installed with new aircraft-quality cap screw bolts (180,000 psi); **$556.33**
- New Ertel Manufacturing Company hardened 30780 intake and 80810 exhaust valve seat inserts installed; Serdi-machined multiangle blueprint valve grind performed; $517
- Intake and exhaust ports fully ported and polished and gasket matched; **$216.67**

Engine Block Total: $4,216.39

The valve seat inserts are installed in the French flathead engine block after they have been Serdi-machined. This is a beautiful job of port blending, also known as pocket porting.

CHAPTER 12
CRANKSHAFT

Part A
Purchase

There appears to be very little information published about the French flathead V-8 crankshaft other than that it is available in a 3 3/4-inch or 4-inch stroke. The crankshafts are made of cast steel.

A 1949–1953 Mercury 6303 cast alloy steel crankshaft with 4-inch stroke was used in the French flathead V-8 engine described in this section. In most instances, the only way to determine whether a crankshaft is from a Mercury flathead V-8 engine is to measure the stroke in the engine block. If the crankshaft is out of the engine block, have an engine building shop measure the stroke. (Measuring the stroke on a garage floor with calipers might provide an incorrect reading.) Some of the Canadian Mercury crankshafts have a small oval indent (jellybean) approximately 3/4 inch long by 3/8 inch wide in the front of the forward counterweight, and some have casting number CM or 61402 on one of the counterweights. Some Mercury crankshafts have casting numbers D2613, F2511, or K0308 on the side of one of the counterweights. The *Ford Motor Company of Canada Master Repair Manual* specifies the out-of-round service limit at 0.0015 inch, the taper service limit at 0.001 inch, and the maximum runout at 0.002 inch.

Magnaflux

Refer to Section I

Shot Peen and Plug Removal

Refer to Section I

Straighten, Chamfer, and Polish

Refer to Section I

Regrinding and Cleaning

The connecting rod journals were reduced in diameter to 2.00 inches in order to accommodate the SCAT Enterprises 2-239-7000-2000 H-beam connecting rods. The 4-inch stroke combined with the 3.2275-inch bore results in a displacement of 261.9 ci.

After all the procedures mentioned in this chapter have been carried out, thoroughly clean the crankshaft with hot water and soap. Use long, thin engine brushes to clean out the oil holes in the main bearing and connecting rod journals. The term "surgically clean" applies here.

This Mercury crankshaft with Federal Mogul 533-M-10 main bearings (0.010-inch oversize) will be installed in the French flathead V-8 engine. All the necessary machine work on the crankshaft has been carried out.

Main Bearings

The 1946–1948 Ford flathead V-8 main bearings are required for the French flathead V-8 engine block. The notch in the rear main bearing saddle for the main bearing shell locking tang (tab) in the French flathead is quite wide, similar to the Model 59 engine block. The notch in the rear main bearing saddle for the main bearing shell in the 1949–1953 Ford flathead is approximately half the width of the Model 59 notch. A new set of Federal Mogul 533-M-10 main bearings (0.010-inch oversize) was installed in the French flathead engine. Refer to Chapter 2 for information on main bearing availability and clearances.

Crankshaft Gear and Pulley

The crankshaft gear supplied with the French flathead V-8 engine is the same rotation as the 1946–1948 Ford

Install a set of Federal Mogul 533-M standard-size main bearings in a 1946–1948 Ford flathead V-8 Model 59 engine.

flathead V-8 crankshaft gear. An aftermarket 1949–1953 distributor was installed in the French flathead engine, requiring the 1949–1953 Ford flathead V-8 timing gears. If the Ford 6303 cast-iron helical crankshaft gear with 22 teeth is missing, replace it with a Melling 2701 crankshaft gear (1949–1953). Melling products are available at most automotive parts outlets.

The stock French flathead crankshaft pulley has four V-belt grooves. This pulley was not used, because a crankshaft pulley with a single groove was required and a blower belt pulley was mounted on the front of the crankshaft pulley. Ford Motor Company never offered a crankshaft pulley for this application. Luke's Custom Machine & Design fabricates a one-piece steel crankshaft hub and single V-belt pulley for Ford flathead V-8 engines equipped with a blower.

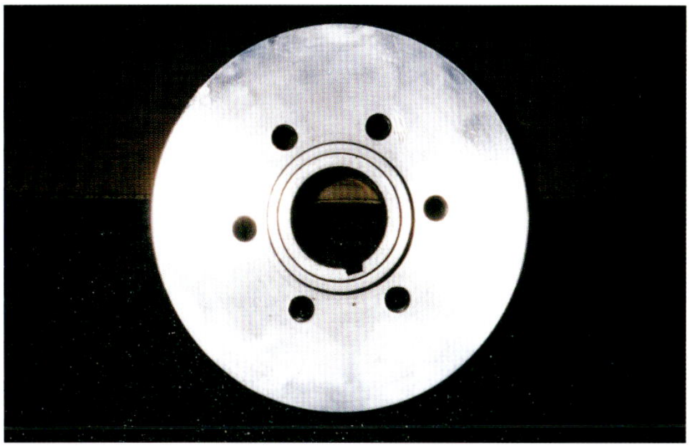

Luke's Custom Machine & Design fabricates a one-piece steel crankshaft hub and single V-belt pulley for Ford flathead V-8 engines equipped with a blower.

Part B

- 1949–1953 Mercury 6303 cast alloy steel crankshaft (Model CM), casting number: 61402; main journal diameter: 2.498 to 2.499 inches; connecting rod journal diameter: 2.138 to 2.139 inches; nonfloating connecting rod bearings; stock stroke: 4.00 inches; crankshaft weight: 67 pounds; **$400**
- Crankshaft Magnafluxed, aligned, shot peened; main journals ground 0.010-inch undersize; connecting rod journals reduced to 2.00 inches diameter; oil holes chamfered; journals polished; crankshaft balanced; factory crankshaft connecting rod journal oil plugs removed and new stainless-steel 1/4-inch NPT plugs installed using Loctite; **$325.62**
- New Federal Mogul 533-M-10 main bearings (0.010-inch oversize) installed allowing for 0.002-inch crankshaft clearance and 0.004-inch endplay; new aircraft-quality main bearing cap bolts (180,000 psi) installed and torqued to 105 ft-lb using molykote; **$120**
- 1949–1953 Ford 6303 cast-iron helical crankshaft gear (22 teeth) and Ford 6310 oil slinger installed with Ford 357654-S crankshaft gear Woodruff key; Luke's Custom Machine & Design new crankshaft sleeve and one-piece crankshaft hub and pulley (6 1/2 inches diameter, single V-belt) installed with Ford 74153-S crankshaft pulley woodruff key and Ford 351590-S crankshaft pulley bolt washer with Ford 20639-S crankshaft pulley bolt; crankshaft pulley bolt installed using Loctite and torqued to 50 ft-lb; oil slinger painted with Glyptal G-1228A medium grey gloss enamel; crankshaft pulley painted with PPG DGHS 9000 black high-gloss polyurethane; **$320.84**

Crankshaft Total: $1,166.46

CHAPTER 13
CONNECTING RODS AND PISTONS

Part A
Purchase

The forged-steel connecting rods supplied with the French flathead V-8 engine are said to be NOS and retail for $225 for a set of eight. These connecting rods are 7.00 inches long (3 3/4-inch stroke) or 6.875 inches long (4-inch stroke) and are similar to the 1949–1953 Ford 6200 Model 8BA and 8BA6205A connecting rods. While that might seem like a bargain for new connecting rods, there are some items that must be added to the initial cost of those connecting rods. The rods should be Magnafluxed, aligned (if necessary), and shot peened. They will also have to be wrist pin–fitted, resized (if necessary), and balanced. This additional machining and labor increases the cost of those connecting rods considerably, and the same procedures must be performed on a set of used 1949–1953 Ford 6200 Model 8BA and 8BA6205A connecting rods. Used Ford flathead V-8 connecting rods are available in abundance and are inexpensive.

Strengthened connecting rods were necessary because the French flathead V-8 engine was equipped with a blower. It is less expensive to purchase a new set of SCAT Enterprises 2-239-7000-2000 H-beam connecting rods than to box a set of Ford 6200 Model 8BA or 8BA6205A connecting rods, and SCAT Enterprises H-beam connecting rods are far stronger than boxed 1949–1953 Ford flathead V-8 connecting rods. The SCAT Enterprises 2-239-7000-2000 connecting rods are manufactured in the United States using 4340 forged steel. They are 7.00 inches long (center to center), and the diameter of the crankshaft end (big end) is 2.1252 inches. A dowel alignment sleeve was fitted to ensure precise connecting rod cap alignment, and the connecting rods were Magnafluxed, shot peened, and stress relieved. The crankshaft connecting rod journals were reduced in diameter to 2.00 inches in order to accommodate the H-beam connecting rods. A proficient crankshaft grinding company can handle that job with ease. ARP 12-point connecting rod bolts are manufactured using 8740 steel at 3/8-inch diameter and 190,000 psi. The average weight of one SCAT Enterprises 2-239-7000-2000 connecting rod is 660 grams, and a set of eight retails for $595.

If your flathead engine will be used for serious competition purposes, upgrade the connecting rod bolts in the SCAT Enterprises H-beam connecting rods with ARP Wave Loc connecting rod bolts, which are manufactured using ARP2000 alloy rated at 220,000 psi. SCAT Enterprises advises these H-beam connecting rods with the upgraded bolts will safely handle 700 to 800 horsepower. Now, that is a flathead!

These are original-design SCAT Enterprises 2-239-7000-2000 H-beam connecting rods. Reduce the connecting rod crankshaft journals to 2.00-inch diameter to accommodate the connecting rods.

Magnaflux and Shot Peen

SCAT Enterprises H-beam connecting rods are Magnafluxed, shot peened, and stress relieved at the factory.

Align, Resize, and Rebush

SCAT Enterprises H-beam connecting rods are aligned, resized, and rebushed at the factory, making the price of these connecting rods even more appealing.

Debeam

SCAT Enterprises H-beam connecting rods do not need to be debeamed.

Connecting Rod Bearings

SCAT Enterprises 2-239-7000-2000 H-beam connecting rods require Clevite 77 CB-610P or Federal Mogul 8-2500RAA standard-size nonfloating connecting rod bearings. If the connecting rod journals are slightly less than the specified 2.00-inch diameter, the Clevite 77 and Federal Mogul connecting rod bearings are also available in 0.010-inch, 0.020-inch, 0.030-inch, and 0.040-inch oversize. Oversize bearings are beneficial for a flathead owner using a 4 1/8-inch (4.125-inch) stroke Mercury crankshaft (with H-beam connecting rods) when the connecting rod journals need to be reground. These bearings are designed for the small-block Buick V-6 and V-8 engines (and other applications), and they cost less than the Ford flathead V-8 nonfloating connecting rod bearings.

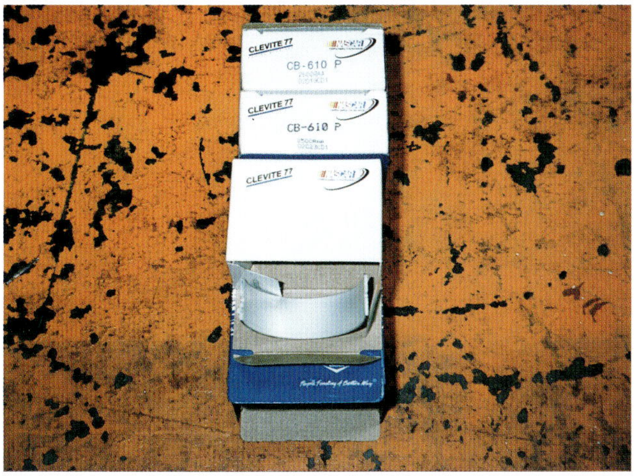

We used a set of Clevite 77 CB-610P standard-size connecting rod bearings with the SCAT Enterprises 2-239-7000-2000 H-beam connecting rods.

If you order a SCAT Enterprises cast-steel crankshaft with 1/8-inch (0.125-inch) connecting rod journal radius, the connecting rod bearings must be chamfered in order to be compatible with the radius. Bearings are chamfered in a lathe using a special jig to hold the bearings. This is a job for only a competent machine shop.

Because original-design SCAT Enterprises H-beam connecting rods were installed, the bottom of the cylinder bores had to be notched for clearance. Due to the fact that the crankshaft is offset in a flathead engine block, the bottom of the cylinder bores next to the camshaft on the left (driver's) side and the bottom of the cylinder bores next to the oil pan rails on the right (passenger's) side had to be notched. Have the cylinder bores notched by a qualified machine shop. The latest design of SCAT Enterprises H-beam connecting rods has been reprofiled, and as a result, notching the bottom of the cylinder bores is no longer necessary. The reprofiled connecting rods weigh 573 grams each.

The notches in the bottom of the cylinder bores allow for clearance with original SCAT Enterprises 2-239-7000-2000 H-beam connecting rods. This is a Crower Cams & Equipment Company blower grind camshaft.

Pistons

The French flathead V-8 pistons are manufactured with cast aluminum for a bore of 3 3/16 inches (3.1875 inches). The same piston is used with the two lengths of French flathead V-8 connecting rods. The current retail price for a set of eight of these pistons with piston rings and wrist pins is $175.

If a Ford flathead V-8 engine will be equipped with a blower, use forged aluminum blower pistons. Blower pistons have a thicker crown area than normally aspirated forged aluminum pistons in order to handle the higher operating temperatures and cylinder pressure of a blown engine. The top compression ring should be at least 0.250 inch below the top of the piston.

A custom set of Ross Racing Pistons forged 2618 T-61 aluminum blower pistons (with a 3.2275-inch bore, 4.00-inch stroke, and 8.0:1 compression ratio) was installed in the French flathead V-8 engine. These pistons have a three-ring slipper skirt design and weigh 351 grams each. The wrist pins are full-floating chrome-moly taper-wall design, weighing 81 grams each, and double Spiro-Lox retainers are supplied with the pistons. The current retail price for the

We used Ross Racing Pistons blower pistons and a Sealed Power E-358K moly-top piston ring set in the French flathead V-8 engine. Notice the large amount of material used for the wrist pin bosses.

Ross Racing Pistons blower pistons, taper-wall wrist pins, and double Spiro-Lox retainers is $830.

Use a moly top piston ring set in an engine equipped with a blower to accommodate the higher operating temperatures and boost. Moly piston rings are much more flexible than cast-iron piston rings. If the second compression ring gap is smaller than the top compression ring gap, there will be less blowby. However, this could cause the top piston ring to flutter. Ross Racing Pistons recommends a larger second compression ring gap than the top piston ring gap in order to prevent piston ring flutter. Ross Racing Pistons recommends a piston ring gap for its moly piston ring sets of 0.004 inch times the bore for the top compression ring plus 0.004 inch for blown engines, and 0.005 inch times the bore for the second compression ring plus 0.004 inch for blown engines. The piston ring gap for the French flathead V-8 engine described in this section will be 0.017 inch top ring, 0.020 inch second ring, and 0.016 inch for the oil ring.

Ross Racing Pistons now sells its off-the-shelf stocking pistons with moly top ring sets as well as the regular Grant cast-iron piston ring sets. The pistons covered by this are: 3 5/16-inch (3.3125-inch), 3 5/16-inch plus 0.030 inch (3.3425-inch), and 3 3/8-inch (3.375-inch) bore in conjunction with a 3 3/4-inch (3.75-inch), 4-inch, or 4 1/8-inch (4.125-inch) stroke. A moly piston ring set will add approximately $50 to the cost of the piston sets.

Balancing

Refer to Section I

Gaskets

Red's Headers sells a complete gasket set for the French flathead V-8 engine for a current retail price of $50. You could purchase the Fel-Pro FS7548 complete gasket set for the 1946–1948 Ford flathead V-8 Model 59 engine, although only the oil pump idler gear cover gasket, water pump gaskets, oil pan gaskets and rear oil seals, timing gear cover gasket, and intake manifold gasket would be used. Copper-sandwiched cylinder head gaskets were installed and are not included in the Fel-Pro FS7548 complete gasket set. It would be more economical to purchase the individual gaskets from Jitney Auto Parts, which provides gaskets for all models of the Ford flathead V-8 engine.

Part B

- New SCAT Enterprises 2-239-7000-2000 forged 4340 steel H-beam connecting rods Magnafluxed, shot peened, stress relieved, and balanced; connecting rod caps fitted with dowel alignment sleeves; new ARP 8740 steel 3/8-inch-diameter, 190,000-psi, 12-point cap screw bolts installed using molykote and torqued to 42 ft-lb allowing for 0.0061 inch stretch; connecting rod weight: 660 grams each; connecting rod length: 7.00 inches; connecting rod ratio: 1.75 (with 4.0-inch stroke); **$595**
- New Clevite 77 CB-610P connecting rod bearings (standard size) installed allowing for 0.002-inch connecting rod clearance and 0.014- to 0.018-inch side clearance per pair of connecting rods; **$48.61**
- New Ross Racing Pistons forged 2618 T-61 aluminum slipper skirt blower pistons; 8.0:1 compression ratio; 3.2275-inch bore (0.040-inch oversize); piston weight: 351 grams each; new Ross Racing Pistons 750-04-28 full-floating heat-treated and case-hardened 4340 chrome-moly steel taper-wall wrist pins installed allowing for 0.002-inch connecting rod clearance; new Ross Racing Pistons 4004-J double Spiro-Lox retainers installed; wrist pin diameter: 0.750 inch; wrist pin length: 2.770 inches; wrist pin weight: 81 grams; piston grooves: top compression ring width: 1.5 millimeters, second compression ring width: 1.5 millimeters, and oil ring width: 4.0 millimeters; new Sealed Power (Federal Mogul) E-358K ductile moly top piston ring set (0.040-inch oversize) installed within manufacturer's recommended arc; top compression ring gap: 0.017 inch, second compression ring gap: 0.020 inch, and oil ring gap: 0.016 inch; new displacement: 261.9 ci (4.3 liters); **$830**
- Complete V-8 engine balance; new French flathead V-8 gasket set installed; **$240**

Connecting Rods and Pistons Total: $1,713.61

CHAPTER 14
LUBRICATION SYSTEM

Part A
Purchase

Little information is available about the French flathead V-8 oil pump. Building a blown French flathead will be expensive, and it is best not to gamble an unfamiliar commodity such as this oil pump that retails for $65. Instead, we installed a new Melling M19 standard-volume oil pump.

Oil Pan and Pickup

The French flathead V 8's die cast aluminum oil pan is similar in design to the 1946–1948 Ford flathead V-8 Model 59 oil pan, except that its sump is deeper at the front of the oil pan. The French flathead oil pan would not be suitable for most vehicles, as the front sump would interfere with the frame crossmember.

The 1946–1948 Ford flathead V-8 Model 59 oil pan fits the French flathead V-8 engine block, provided the forward section of the oil pan is modified to clear the front main bearing cap. An experienced machine shop can handle this project with ease. Block off the crankcase vent tube at the right (passenger's) side front of the Model 59 oil pan. We installed a NOS Canadian Ford Military 6675 oil pan, designed for 1942–1948 flatheads, on the French flathead V-8 engine. Refer to Section I for a description of this oil pan and more information about oil pans. Using a Model 59 oil pan with a Melling M19 standard-volume oil pump dictates that a 1949–1953 Ford Truck 6615 oil pump cover and inlet tube (oil pump pickup) be installed.

If the rear upper oil seal retainer for the French flathead V-8 engine block is missing, use a 1946–1948 Ford 91A-6335 rear upper oil seal retainer. This item is readily available at minimal cost. It may be necessary to mill approximately 0.002 inch off the sides of the lip of the Ford 91A-6335 oil seal retainer in order to properly fit it into the French flathead V-8 engine block.

Oil Pump Idler Gear

The oil pump idler gear and cover in the French flathead V-8 engine are similar to the 1946–1948 Ford flathead V-8 Model 59 oil pump idler gear and cover. If the French oil pump idler gear and cover are not available, the Model 59 oil pump idler gear and cover are easy to locate and inexpensive.

Fuel Pump Pushrod Bushing

As previously mentioned, the fuel pump pushrod bushing in the French flathead V-8 engine is threaded in place. The engine described in this section was equipped with a Holley 600-cfm carburetor and an electric fuel pump. The fuel pump pushrod bushing must be blocked off to prevent a loss of oil pressure. The fuel pump pushrod bushing was left in place, and a slightly tapered 0.460-inch-diameter aluminum plug was installed using Permatex aviation form-a-gasket.

Baffles and Breather Tube

The two valve lifter gallery oil-return hole baffles and the crankcase breather tube are usually not provided with a bare French flathead V-8 engine block. It is a good practice

This 1949–1951 Mercury 6615 oil pump cover and inlet tube with attached oil baffle have been sprayed with graphite.

to install these items, however, as they will prevent debris from falling into the oil pan. Ford 6524 valve lifter gallery oil-return hole baffles are readily available at little or no cost. The 1949–1953 Ford 6756 valve lifter gallery crankcase breather tube is 1 5/16 inches (1.3125 inches) in diameter and will not fit the French flathead V-8 engine block, which has a 1 1/8-inch (1.125-inch) diameter breather tube opening. Cut and install a piece of steel tubing (1 1/8 inches diameter by 5 inches long). Glass bead and paint these parts with Glyptal G-1228A medium grey gloss enamel prior to installing them.

Oil Filter

There is not sufficient room to mount an oil filter on the left (driver's) side cylinder head when a GMC 4-71 blower is installed. Mount a Trans-Dapt 1045 remote oil filter bracket with the Trans-Dapt 1032 neoprene rubber hose kit on the lower part of the firewall. These items are available from most speed-equipment dealers.

Part B

- New stainless-steel front oil gallery line plug (1/8-inch NPT) installed using pipe thread sealant; fuel pump pushrod bushing blocked off with tapered 0.460-inch-diameter aluminum plug using Permatex aviation form-a-gasket; 1946–1948 Ford 6657 oil pump idler gear bushing installed with Ford 6655 oil pump idler gear and Ford 6656 oil pump idler gear shaft allowing for 0.002-inch clearance; 1946–1948 Ford 13-6658 oil pump idler gear cover installed with new stainless-steel bolts and lock washers using Loctite and torqued to 15 ft-lb; oil pump idler gear cover gasket installed using silicone sealant; inside of oil pump idler gear cover, valve lifter gallery crankcase breather tube, and Ford 6524 valve lifter gallery oil-return hole baffles (2) painted with Glyptal G-1228A medium grey gloss enamel; **$105.66**
- New Melling M19 standard-volume oil pump (for 1949–1953 Ford) installed with 1949–1953 Ford Truck 6615 oil pump cover and inlet tube, NOS Ford 6623 oil pump cover screen, and Ford 6628 oil pump cover screen spring retainer; oil pump end clearance: 0.00025 inch; oil pump cover and inlet tube gasket installed using silicone sealant and bolts torqued to 80 in-lb using Loctite; 1946–1948 Ford 91A-6335 rear upper oil seal retainer installed using silicone sealant; rear oil seal packings (Ford 6700) soaked in SAE 30-weight motor oil and installed; new Speedway Motors 912-S12853 front one-piece oil seal installed using silicone sealant; **$190.35**
- NOS 1942–1948 Canadian Ford Military 6675 oil pan (4-quart capacity) with removable bell housing section installed with new stainless-steel bolts, lock washers, and AN flat washers using Loctite and torqued to 15 ft-lb; oil pan gaskets installed using silicone sealant; Ford 6754 dip stick tube and Ford 6750 dipstick installed; oil pan, dip stick tube, and dipstick Redi-Stripped and painted with PPG 74000 red high-gloss polyurethane; engine lubricated with 5 quarts Pennzoil HD-30-weight motor oil; **$429.41**
- New Trans-Dapt 1045 polished remote oil filter bracket with horizontal inlet/outlet; new Morosso 22400 chrome oil filter; new Trans-Dapt 1032 neoprene rubber hoses (2 feet long by 1/2-inch NPT) installed; hoses covered with new Spectre 3001A black nylabraid and new Spectre 3308B chrome champclamps; new Weatherhead 48-8 polished-brass compression fittings (1/2-inch tube to 3/8-inch NPT); new Weatherhead 3328-6 polished-brass 3-inch male extension (3/8-inch NPT); new Weatherhead 3300-6 polished-brass female coupling (3/8-inch NPT); new Weatherhead 3400-6 polished-brass 90-degree elbow (3/8-inch NPT); and new Weatherhead 3220-6-2 polished-brass reducer (1/8-inch NPT to 1/4-inch NPT) installed using pipe thread sealant; **$96.94**

Lubrication System Total: $822.36

CHAPTER 15
CAMSHAFT AND CYLINDER HEADS

Part A
Camshaft Bearings

The French flathead V-8 engine blocks have the camshaft bearings installed. Have the camshaft bearings removed when hot tanking the engine block in order to dissolve the protective coating applied to the engine block. Once camshaft bearings are removed from an engine, they should not be reused, as the camshaft bearings will likely be damaged when they are removed. There does not appear to be any information available regarding the manufacturer of the camshaft bearings and the quality of the bearings. The expense of a blown French flathead V-8 engine dictates that only quality engine parts be used. I recommend installing new Clevite 77 SH-21S or new Federal Mogul 1005M camshaft bearings.

The custom-ground camshaft supplied by Crower Cams & Equipment Company is an excellent street camshaft for the blown French flathead V-8 engine.

Camshaft

There is little information available about the French flathead V-8 camshaft. It is similar to a 1949–1953 Ford flathead V-8 camshaft with the distributor gear on the nose, and the lift is said to be 0.316 inch. Whatever the specifications are, the camshaft would not be suitable for a street-performance flathead, especially a blown one. The current retail price for one of these camshafts is $25.

The specifications for the camshaft in a blown street-performance engine are different from the specifications for a normally aspirated engine. As far as I know, none of the major camshaft manufacturers in the United States sell an off-the-shelf blower camshaft for a Ford flathead V-8 engine. This means one will have to be ground, which is a good thing—the camshaft will be perfectly suited for your specific engine. A street-performance blower camshaft should have a large lobe separation angle, from 112 to 114 degrees. A camshaft with long duration and big overlap is not recommended.

Crower Cams & Equipment Company recommends the SPL8BA blower grind custom flat tappet (lifter) camshaft for the French flathead V-8 engine. This camshaft has an advertised duration of 277 degrees intake and exhaust; duration at 0.050-inch lift: 224 degrees intake and exhaust; lobe separation angle: 112 degrees; net valve lift: 0.321-inch intake and 0.319-inch exhaust; and valve lash (hot): 0.010-inch intake and 0.012-inch exhaust. This is a terrific street blower camshaft, and the price is outright with no core exchange.

Camshaft Interchange

Refer to Section I

Degree Camshaft

Refer to Section I

The Schneider Racing Cams 254-F solid lifter camshaft is ideal for the smaller-cubic-inch 1949–1953 Ford flathead V-8 engines.

Valve Lifters

The valve lifters (or tappets or cam followers) in the French flathead V-8 engine are taller than those used in the 1946–1953 Ford flathead V-8 engines. As a result, shorter valves are required, which prohibit the use of Ford flathead lifters with French flathead valves or Ford flathead valves with French flathead lifters. French flathead lifters are available in two styles: hollow-body nonadjustable and hollow-body adjustable. The current retail price for a set (16) of hollow-body nonadjustable lifters is $25, and the hollow-body adjustable lifters are $140 for a set (16).

We installed new stainless-steel 4.90-inch-diameter intake and exhaust valves. These valves are longer than the French flathead valves, meaning the French flathead valve lifters cannot be used. Instead, we used new Iskenderian Racing Cams F85-02 solid-body Johnson adjustable lifters. A few decades ago, Tyco Automotive Corporation (New York) used to manufacture solid-body adjustable lifters for the Ford flathead V-8 engines. The indents in the side of those valve lifters was a square shape, unlike the Iskenderian Racing Cams F85-02, Clevite 213-1614, Sealed Power AT770, TRW VL36N, and Perfect Circle 231-1614 solid-body valve lifters, which all have oval-shaped indents.

Valves

There is little information available about the French flathead V-8 intake and exhaust valves. It is known, however, that they have a 1.50-inch head diameter and 11/32-inch valve stem diameter. As previously mentioned, they are shorter than the Ford flathead V-8 valves and will not be used. The current retail price for French flathead valves is $5 each or $80 for a set of 16. It is interesting to note that the French flathead valves were fitted with a valve stem seal for oil-control purposes.

We installed new Motor City Flathead P2091HP-C stainless-steel intake valves (1.72-inch head diameter, 11/32-inch stem diameter, 4.90-inch length) and new P2323JF stainless-steel exhaust valves (1.50-inch head diameter, 11/32-inch stem diameter, 4.90-inch length) with a Serdi-machined multiangle blueprint valve grind. These valves had been previously supplied with the engine block. For a blown street-performance flathead engine, it would be better to install 1.61-inch head diameter valves for both the intake and exhaust.

Valve Guide Bushings

The one-piece valve guide bushings used in the French flathead V-8 engine are similar in design to the one-piece valves

Motor City Flathead P2091HP-C stainless-steel 1.72-inch-diameter intake valves and P2323JF stainless-steel 1.50-inch-diameter exhaust valves were used in the French flathead V-8 engine.

guide bushings used in the 1949–1953 Ford flathead V-8 engine. The French flathead V-8 valve guide bushings were also equipped with O-ring seals for oil control. It is unknown if the French flathead V-8 valve guide bushings are the same 1.030-inch o.d. and 2.20-inch overall length as the 1949–1953 Ford flathead V-8 valve guide bushings. To be safe, we installed NOS 1949–1953 Ford 6510 one-piece valve guide bushings.

Valve Springs

Because the valves are shorter in the French flathead V-8 engine, the valve springs are also shorter. There seem to be no specifications available for those valve springs other than the current retail price, which is $30 for a set (16).

Instead of using the French flathead valve springs, we installed a new set of Iskenderian Racings Cams 4005 dual valve springs. These valve springs have sufficient spring tension for use in a blown street-performance engine. Iskenderian Racing Cams 185-G single valve springs would also be appropriate, as they have adequate spring tension for the intake valves in the closed position to prevent the intake valves from being pulled off the valve seat inserts when the engine is under boost. Installing single valve springs rather than dual valve springs will save a few hours of labor. However, if you want to sweat the way you would in a Finnish sauna, install the dual valve springs. A new set of Iskenderian Racing Cams 4005 dual valve springs was measured at High Performance Engines using a Rinck-McIlwane precision valve spring tester. The results were:

Installed height, valves closed: 60 pounds at 1.90625 inches (1 29/32 inches), or 82 pounds at 1.8125 inches, or 115 pounds at 1.6875 inches

Installed height, valves open: 152 pounds at 1.5365 inches, or 163 pounds at 1.50625 inches, or 170 pounds at 1.4625 inches (0.350-inch lift)

1.240 inches o.d. and 0.925-inch/0.690-inch i.d.

Coil bind: 1.345 inches

Color: grey

We measured an Iskenderian Racing Cams 4005 dual valve spring on a Rinck-McIlwane precision valve spring tester.

Valve Spring Retainers and Locks

A set of Iskenderian Racing Cams 87-F dual valve spring retainers and a good used set (16 pieces) of 1952–1953 Ford 6514 valve spring retainers and 6517 sleeves was glass beaded and installed in the French flathead V-8 engine. A new set of Manley Performance Products 13127-16 stamped-steel valve stem locks was included as well. A good used set (16 pieces) of 1949–1953 Ford 6512 crow's foot valve guide bushing retainers was glass beaded and installed.

Camshaft Gear and Timing Cover

The camshaft gear installed in the French flathead V-8 is manufactured out of aluminum; however, it should not be

This picture shows Iskenderian Racing Cams 87-F dual valve spring retainers and VL-11/32-inch valve stem locks, 1952–1953 Ford 6514 valve spring retainers, and Ford 6512 crow's foot valve guide bushing retainers.

used with a 1949–1953 Ford flathead V-8 camshaft, because this camshaft gear provides thrust in the opposite direction as the 1949–1953 Ford camshaft gear. The current retail price for a French flathead aluminum camshaft gear is $50.

Replace the French flathead V-8 camshaft and crankshaft gears with a new Melling 2727S or Federal Mogul 221-2727S matched aluminum camshaft gear and cast alloy steel crankshaft gear set.

The timing gear cover used on the French flathead V-8 engine is similar in design to the 1949–1950 Ford flathead V-8 cast-iron timing gear cover, except that it has an adjustable bolt in the center to act as a camshaft thrust button. Ford Motor Company used to offer a front cover kit for the 1949–1950 Ford flathead V-8 engine. The Ford 1284 timing gear cover has an adjustable camshaft thrust bearing surface to control the camshaft end movement. It is highly doubtful that any of those kits still exist today.

An MSD 8574 electronic distributor was used with the French flathead V-8 engine, and this required the installation of a 1949–1950 Ford 8BA-6059 cast-iron timing gear cover. Refer to Chapter 7 for more information on timing gear covers and distributors.

One-Piece Oil Seal

Refer to Section I

Cylinder Heads

The cylinder heads on the French flathead V-8 engine were manufactured using cast iron. They are similar in design to the 1949–1953 Ford flathead V-8 cylinder heads with the cast-iron water outlet elbow at the top front of each cylinder head. There may be a casting number of 701643 on a left cylinder head and casting numbers 703146 and 710084 on right cylinder heads. The 1939–1953 Ford flathead V-8 or aftermarket aluminum cylinder heads will bolt directly onto the French flathead engine block without any modifications. We installed a pair of new Edelbrock 1115 aluminum cylinder heads (1949–1953) on the French flathead V-8 engine block. It will be necessary to unshroud (lay back) the combustion chambers in the Edelbrock cylinder heads above and beside the intake valves in order to provide adequate clearance for the intake valves with 1.72-inch head diameter.

The cylinder head bolt hole threads in the French flathead V-8 engine block are deeper than the cylinder head bolt hole threads in the 1946–1953 Ford flathead V-8 engine blocks, which means the 1949–1953 Ford flathead V-8 cylinder head bolts should not be used. They will not penetrate far enough into the threads to give the proper clamping force. The center bolt hole on each side of the French flathead V-8 engine block is in the exhaust port runner and not the deck surface, thus requiring a longer bolt. The French flathead V-8 bare engine blocks are not supplied with cylinder head bolts and, as a result, new Grade 8 (150,000 psi) cylinder head bolts must be purchased. The bolts required are two pieces of 7/16-inch N.C. by 4 1/4 inches long, 22 pieces of 7/16-inch N.C. by 3 inches long, four pieces of 7/16-inch N.C. by 2 3/4 inches long, and 20

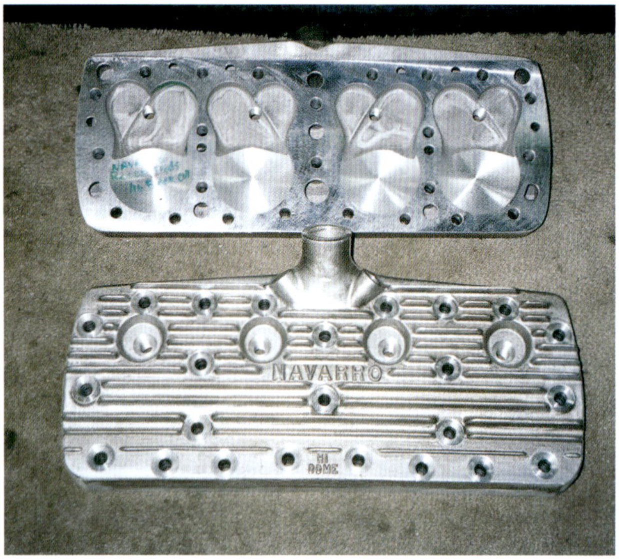

This pair of 1939–1948 (Barney) Navarro aluminum high-dome cylinder heads has a combustion chamber volume of 102 cc.

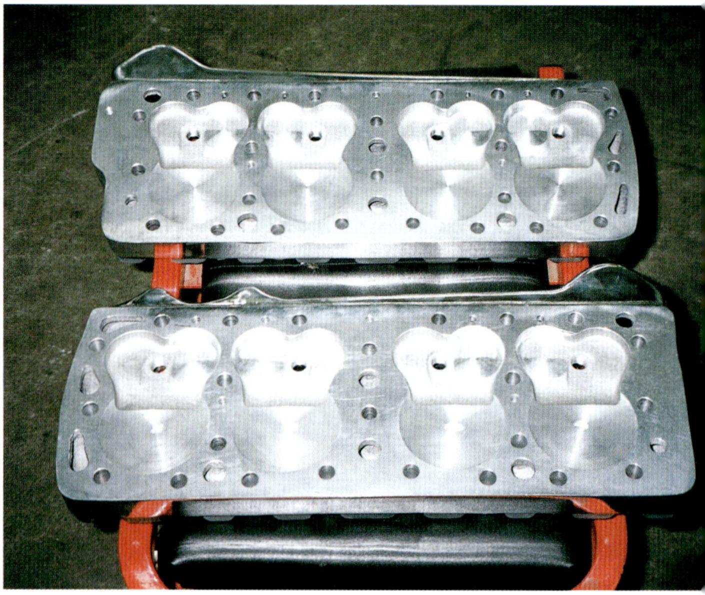

The Edelbrock 1115 aluminum cylinder heads were unshrouded around the intake valves in order to provide adequate valve-to-cylinder-head clearance.

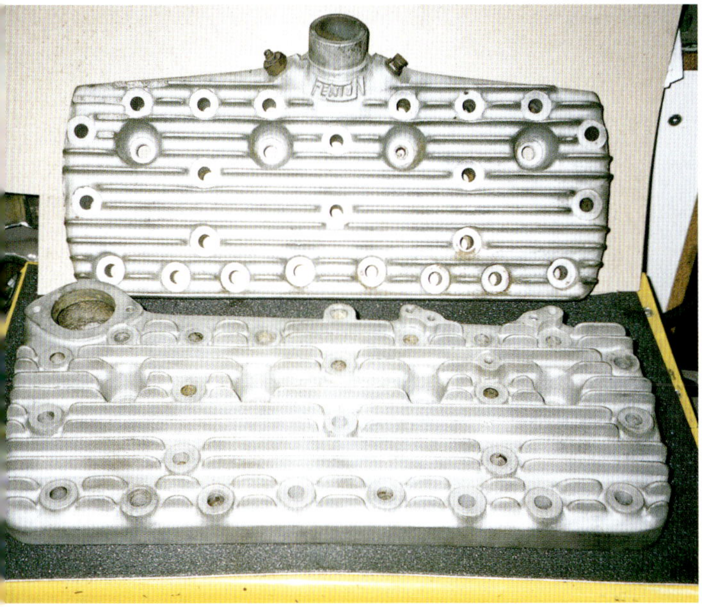

The top vintage aluminum cylinder head is a 1939–1948 Fenton model with a 71-cc combustion chamber. The lower vintage aluminum cylinder head is a 1949–1953 Edmunds model with a 73-cc combustion chamber.

pieces of 7/16-inch N.C. by 2 1/4 inches long. The Grade 8 bolts are readily available at industrial supply stores for approximately $16 for the complete set. Jitney Auto Parts sells chrome acorn nut covers for the 5/8-inch head size of the Grade 8 cylinder head bolts. Refer to Chapter 5 for information pertaining to cylinder head bolt washers.

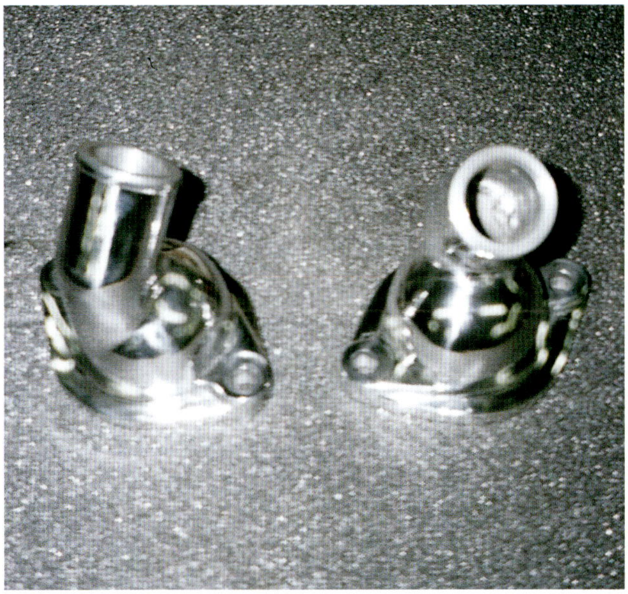

Luke's Custom Machine & Design modified the Offenhauser 3496 water outlet elbows, which were then buffed by "Tim the polishing guy."

Cylinder Head Gaskets

Edelbrock recommends Fel-Pro 1055 (right side) and 1056 (left side) copper-sandwiched cylinder heads gaskets for use with its 1115 aluminum cylinder heads (1949–1953). These gaskets are suitable for cylinder bores up to 3.420 inches, they are 0.062 inch thick, and their compressed volume is 14 cc. Fel-Pro 1055 and 1056 cylinder head gaskets can be purchased from most automotive parts outlets or from Speedway Motors (910-15895).

Water Outlets

The Offenhauser 3496 straight water outlet elbows must be modified to clear the GMC 4-71 blower case. Luke's Custom Machine & Design cuts the radiator hose connection off the water outlet elbow, fabricates a new connection, and welds this on in a new position approximately 45 degrees from the old position.

Exhaust Baffles

The center cylinder head bolt hole (on each side) in the French flathead V-8 engine block (where the stud that secures the exhaust baffle is installed) is not threaded in the deck surface. There is a blind hole through the exhaust runner below the deck surface that is threaded. This blind hole has a bump on the end that acts as a semibaffle in the exhaust port. The fact that this hole in the deck is not threaded is an open invitation for a water leak.

Exhaust baffles should be installed; refer to Section I for a full explanation. The round end of the Speedway Motors 916-15201 exhaust baffle must be milled to a concave shape in order to rest against the bump at the lower side of the blind cylinder head bolt hole, and the exhaust baffle must be reduced in length by one half. If this is not

This is how the Speedway Motors 916-15201 exhaust baffle should look prior to being installed in a French flathead V-8 engine block.

done, the exhaust baffle will protrude far enough down the exhaust port so as to obstruct the exhaust flow. The exhaust baffle is not held in place with the stud, so it must be secured in position with a 1/4-inch N.C. socket head set screw and jam nut using Loctite. Drill a hole through the exhaust port into the exhaust baffle, thread the hole, and install the set screw with the jam nut.

Part B

- New Clevite 77 SH-21S camshaft bearings installed allowing for 0.002-inch camshaft clearance and 0.007-inch camshaft end play; number 1 and 2 bearings: SH-21, housing bore: 1.9275 to 1.9285 inches; number 3 bearing: SH-22, housing bore: 1.9275 to 1.9285 inches; camshaft journal diameter: 1.7965 to 1.7970 inches; **$27.95**
- New Crower Cams & Equipment Company SPL8BA blower grind cast-iron alloy solid-lifter camshaft with advertised duration: 277 degrees intake and exhaust; duration at 0.050-inch lift: 224 degrees intake and exhaust; net valve lift: 0.321-inch intake and 0.319-inch exhaust; lobe separation angle: 112 degrees; valve lash (hot): 0.010-inch intake and 0.012-inch exhaust; camshaft degreed; Ford 6254 rear camshaft gear (camshaft to oil pump idler gear) and 1949–1950 Ford 7RA-6254-B front camshaft gear (camshaft to distributor drive gear) installed; **$275.01**
- New Iskenderian Racing Cams F85-02 Johnson hard-enable iron solid adjustable lifters (stock 1.00-inch diameter) installed; solid-body lifter weight: 114 grams each; **$139.95**
- New Iskenderian Racing Cams 4005 aircraft-quality racing dual helical coil valve springs (oil tempered, shot peened, made of silicon chrome); installed height: 82 pounds at 1.8125 inches valves closed and 170 pounds at 1.4625 inches (0.350-inch lift) valves open; coil bind: 1.345 inches; 1.240-inch o.d. and 0.925-inch/0.690-inch i.d.; color: grey; new Iskenderian Racing Cams 87-F dual valve spring retainers and 1952–1953 Ford 6514 valve spring retainers and 6517 sleeves installed with new Manley Performance Products 13127-16 standard valve stem locks (7 degree, 4140 chrome-moly stamped steel, heat-treated, 11/32-inch stem diameter); NOS Ford 6510 one-piece valve guide bushings installed with 0.002-inch valve guide boss clearance and 0.002-inch valve stem clearance; 1949–1953 Ford 6512 valve guide bushing retainers installed; **$239.02**
- New Motor City Flathead P2091HP-C stainless-steel 1.72-inch head diameter intake valves (hardened tips, swirl polished, 11/32-inch stem diameter, 4.90 inches long overall) and new Motor City Flathead P2323JF stainless-steel 1.50-inch head diameter exhaust valves (undercut stems at head, hardened tips, swirl polished, 11/32-inch stem diameter, 4.90 inches long overall) installed; **$159.20**
- New Melling 2702 aluminum camshaft helical timing gear (1949–1953) with 44 teeth installed with Ford 6258 lock washer plate and Ford 350400 camshaft gear bolts using Loctite and torqued to 20 ft-lb; 1949–1950 Ford 8BA-6059 chrome cast-iron timing gear cover installed with new polished stainless-steel hex head bolts and lock washers using Loctite and torqued to 15 ft-lb; timing gear cover gasket installed using silicone sealant; **$169.92**
- New Speedway Motors 916-15201 exhaust baffles installed; **$34.95**
- New Fel-Pro 1055 and 1056 copper-sandwiched cylinder head gaskets (0.062-inch thickness, 14-cc compressed volume) installed using Permatex copper spray-a-gasket; new Edelbrock 1115 polished 356 T-6 finned aluminum cylinder heads (8.0:1 compression ratio, 63-cc combustion chamber volume) installed with new Grade 8 cylinder head bolts (150,000 psi) and new Manley Performance Parts 42102 heat-treated head bolt washers (7/16-inch i.d.) using Permatex aviation form-a-gasket and torqued to 40 ft-lb; new J.C. Whitney & Company 38BB2776R chrome acorn nut covers (5/8-inch head size) installed; valve-to-cylinder-head clearance: 0.185-inch intake and 0.195-inch exhaust (without gasket); **$740**
- New Offenhauser 3496 straight polished-aluminum water outlet elbows modified by Luke's Custom Machine & Design and installed with new polished stainless-steel hex head bolts and lock washers; water outlet elbow gaskets installed using silicone sealant and bolts torqued to 15 ft-lb; new Speedway Motors 910-15802 thermostats (180 degrees Fahrenheit) installed; **$145**

Camshaft and Cylinder Head Total: $1,931

CHAPTER 16
INTAKE SYSTEM

Part A
Intake Manifold

The French military flathead V-8 engine was equipped with a single two-barrel cast-aluminum intake manifold similar to the 1949–1953 Ford flathead V-8 intake manifold. Aftermarket 1939–1953 Ford flathead V-8 intake manifolds will fit the French flathead V-8 engine block without modification other than gasket matching the intake ports.

The original Harrell tri power intake manifold for Ford flathead V-8 engines is a unique two-piece unit, similar to a tunnel ram intake manifold. The tall top section used to mount the carburetors was bolted to the lower section. The first dual-carburetor intake manifolds of this design cast were named "Tornado/Harrell." The "Tornado" name had been used by Lee's Speed Shop (Lee Chapel, Oakland, California) since the 1940s, so the "Tornado" name was dropped and the intake manifold was renamed "Harrell." Floyd Hulegaard (Eugene, Oregon) is now having a reproduction of the tri-power lower section cast with high-grade aluminum for use as a blower intake manifold. The raw casting of this blower intake manifold does not have a flange on the top to mount the blower; this flange had to be fabricated. Remove the plenum divider by milling or grinding, and install a pop-off valve. Use the intake manifold gasket used as a template to scribe around the intake ports in the engine block as a template for the intake manifold to gasket match the ports. Grind off any

Luke's Custom Machine & Design completed this Harrell blower intake manifold. The pop-off valve was fabricated using an overhead valve (OHV) Ford 2-inch-diameter intake valve.

Use a (Barney) Navarro blower intake manifold to mount a GMC 4-71 blower on a flathead engine. Luke's Custom Machine & Design modified this intake manifold to accept a GMC 6-71 blower for use at the Bonneville Salt Flats.

Luke's Custom Machine & Design fabricated this finned-aluminum oil separator. This is a practical addition for any engine with a PCV valve that is not located over an oil baffle.

casting flash in the ports and plenum. Use a piece of quality gasket material to fabricate the intake-manifold-to-blower gasket.

As a point of interest, Ken Austin (see Appendix B: Resources) will soon offer a blower intake manifold for Ford flathead V-8 engines. This intake manifold will have a pop-off valve, a tented floor in the plenum, and it will certainly be of the same premium quality as all of his Ardun intake manifolds.

The Harrell cast-aluminum blower intake manifold does not have holes in the left front for the oil filler/breather tube and road-draft/ breather tube. As a result, we installed a 1939–1948 Ford 9415 fuel pump stand/crankcase breather housing. Jitney Auto Parts markets a polished cast-aluminum finned crankcase breather/oil filler housing as an alternative to the Ford flathead V-8 model.

We installed a PCV valve in the Harrell blower intake manifold. If the PCV valve is not located over an oil baffle, install an oil separator between the PCV valve and the carburetor to prevent motor oil from being sucked into the carburetor. Luke's Custom Machine & Design fabricated a compact aluminum finned oil separator that will be bolted to the rear of the blower case on the French flathead V-8 engine. Also refer to Chapter 6.

Here is a new old stock (NOS) original Italmeccanica blower with a three-belt drive pulley, a dual-carburetor adapter cast into the top of the blower case, and a 1949–1953 Ford flathead V-8 intake manifold with a pop-off valve. This is one of the rarest automotive parts in existence and is literally worth a king's ransom.

wizards might argue that a 500- or 600-cfm carburetor is far too much for a 262-ci flathead engine. According to the previously mentioned formula, they are theoretically correct. However, it is better to run on the rich side with a blown motor. If the engine ever leans out and backfires, the whole show will be over.

Holley recently introduced a new line of four-barrel carburetors designed specifically for supercharged engines. These nonemission Model 4150 carburetors have dual 50-cc accelerator pumps, an intake manifold referenced power valve, and a manual choke. We installed a Holley 0-80592S silver-finish 600-cfm four-barrel carburetor on the French flathead V-8 engine. The current retail price for this carburetor is $415.

An adapter is required to mount the Holley four-barrel carburetor on the GMC 4-71 blower. The Blower Drive Service 14B4 single four-barrel carburetor adapter is a nice polished unit complete with gaskets, bolts, and a blower inlet safety screen. This item is available for the current retail price of $95.

The Blower Drive Service 14B4 polished-aluminum single four-barrel carburetor adapter for a GMC 4-71 blower.

Fuel Pressure Regulator and Fuel Line

The stock Ford flathead V-8 fuel pump will not supply sufficient fuel pressure for a modern four-barrel carburetor; an electric fuel pump and a fuel pressure regulator must be installed. The Holley 12-803 fuel pressure regulator is adjustable from 4 1/2 to 9 psi, and it has a silver finish.

The Holley 0-80592S carburetor requires the Mr. Gasket 1552 chrome dual-inlet fuel line. Refer to Chapter 6 for information about a fuel-pressure gauge.

Install a vacuum/boost gauge below the blower. The Auto-Meter 4401 Pro Comp Ultra-Lite vacuum/boost gauge is a presentable unit. Holley, Mr. Gasket, and Auto-Meter products are available from most speed-equipment merchants.

Air Cleaners

We installed a Speedway Motors 925-11904 Mr. Roadster polished cast-aluminum finned single-quad Cobra-style air cleaner on the blown French flathead V-8 engine. This air cleaner is 15 1/4 inches long by 8 1/2 inches wide by 3 5/8 inches high, and it looks perfectly compatible with the finned aluminum cylinder heads. We installed a K&N E-3515 air filter (14.625 inches long by 7.75 inches wide by 2.00 inches high) to promote maximum airflow. The current retail price for air cleaner is $130, and the air filter is available for $65.

Bolts

Refer to Section I

Part B

- New Harrell polished-aluminum 1939–1948 Ford flathead V-8 blower intake manifold installed with 1949–1953 Ford 20448-S intake manifold bolts and new stainless-steel AN flat washers using Loctite and torqued to 25 ft-lb; intake manifold gasket installed using silicone sealant; new Speedway Motors 910-10125 chrome acorn nut covers (9/16-inch head) installed; Luke's Custom Machine & Design pop-off valve fabricated with Ford OHV 2-inch-diameter intake valve; new Jitney Auto Parts polished-aluminum fuel pump stand/crankcase breather housing installed with new polished stainless-steel acorn nuts and AN flat washers using antiseize compound; fuel pump stand/crankcase breather housing gasket installed using silicone sealant; new Mr. Gasket 2060 chrome push-in oil filler/breather cap installed; new Mopar Performance (Chrysler Corporation) BU-1412 PCV valve installed with new Aeroquip FBA0800 braided stainless-steel line (3/8-inch i.d.) using new stainless-steel line clamps; Luke's Custom Machine & Design new polished finned-aluminum oil separator installed with new polished stainless-steel bracket and new Weatherhead 1069-6-2 polished-brass 90-degree male elbows (3/8-inch i.d. hose to 1/8-inch NPT) using pipe thread sealant; **$774.89**
- GM 5114200 polished-aluminum large-bore die-cast Model 4-71 (blower) supercharger case; GM 5122363 polished die-cast aluminum front and rear bearing

plates; new Blower Drive Service 671SL1 bearing seals; new MRC 5205 front bearings; new MRC 6205 rear bearings; GM 5114442 polished die-cast aluminum front cover assembled with new polished stainless-steel socket head bolts, lock washers, and AN flat washers using antiseize compound; new Blower Drive Service 471GK gasket kit installed using Permatex aviation form-a-gasket; new Blower Drive Service 555 pressure-relief valve and 556 oil level sight gauge and new stainless-steel oil filler and oil drain plugs (3/8-inch NPT) installed using pipe thread sealant; blower case pruned and shaped; rotors double-pinned and clearanced; front and rear bearing plates machined for bearing support rings; rear polished-aluminum cover fabricated by Luke's Custom Machine & Design; blower installed with new Papco 264-016 engine studs (5/16 inch by 1 3/4 inches long) and new polished stainless-steel AN flat washers and nylocks using antiseize compound; blower gears lubricated with Castrol Syntec 75W-90 gear oil; **$1,830.28**

- Luke's Custom Machine & Design new LS2000 polished 6061 T-6 aluminum sculptured blower drive snout, new LS2200 blower drive snout shaft (4140 chrome-moly heat-treated steel), and new LS4100 blower drive snout coupler (4140 chrome-moly heat-treated steel) installed with new stainless-steel socket head bolts using antiseize compound; **$500**
- Luke's Custom Machine & Design new polished 6061 T-6 aluminum blower belt idler pulley (5-inch o.d.) and new polished-aluminum idler pulley adjustment bracket installed with new polished stainless-steel AN flat washer, lock washer, and nut using antiseize compound; **$518.52**
- New Blower Drive Service 6330 polished-aluminum crankshaft blower drive pulley (30 tooth by 3 1/2 inches wide by 1/2 inch pitch, Gilmer) and new Blower Driver Service 6335 blower snout drive pulley (35 tooth by 3 1/2 inches wide by 1/2 inch pitch, Gilmer) installed with new polished stainless-steel socket head bolts, lock washers, and AN flat washers using antiseize compound and torqued to 25 ft-lb; blower drive pulleys reduced in width to 2 1/2 inches; new Blower Drive Service 560H200 blower drive belt (1/2 inch pitch, Gilmer, by 2 inches wide by 56 inches long) installed; blower drive ratio: 14.2 percent underdrive; boost: 8.24 pounds (estimated); **$265.33**
- New Holley 0-80592S Model 4150 nonemission 600-cfm four-barrel supercharger carburetor (silver finish) installed with new ARP 400-2401 polished stainless-steel carburetor studs (standard length) using antiseize compound; new Blower Drive Service 14B4 polished-aluminum carburetor-to-blower adapter (1 inch thick) installed with new stainless-steel socket head bolts using antiseize compound; new Super Sunny S6052 chrome carburetor return spring installed; **$605.66**
- New Speedway Motors 925-11904 Mr. Roadster polished cast finned-aluminum Cobra-style air cleaner (15 1/4 inches long by 8 1/2 inches wide by 3 5/8 inches high) installed with new polished stainless-steel AN flat washer and acorn nut using antiseize compound; new K&N E-3515 air filter (14.625 inches long by 7.75 inches wide by 2.00 inches high) installed; **$195**
- New Mr. Gasket 1552 chrome dual-inlet fuel line (9 5/16 inches center to center) installed; new Holley 12-803 fuel-pressure regulator (adjustable from 4 1/2 to 9 psi) installed with new Weatherhead 1069-6 polished-brass 90-degree male elbow (3/8-inch i.d. hose to 1/4-inch NPT); new Weatherhead 3320-6-4 polished-brass bushing (3/8-inch NPT to 1/4-inch NPT, inlet line); new Weatherhead 3220-6-2 polished-brass bushing (3/8-inch NPT to 1/8-inch NPT); new Weatherhead 3325-2 brass hex nipple (1/8-inch NPT, outlet line); new Weatherhead 3220-6-2 polished-brass bushing (3/8-inch NPT to 1/8-inch NPT) and new Pro Comp Ultra-Lite 4411 fuel-pressure gauge (1 to 15 psi, 2 5/8-inch-diameter face); all fittings installed using pipe thread sealant; new Auto-Meter 4401 Pro Comp Ultra-Lite vacuum/boost gauge (0 to 30 hg./0 to 20 psi, 2 5/8-inch-diameter face) installed with new Weatherhead 1066-4 polished-brass female connector (1/4-inch i.d. hose to 1/8-inch NPT); **$138.29**

Intake System Total: $4,827.97

CHAPTER 17
IGNITION SYSTEM

Part A
Distributor

The ignition system used with the French flathead V-8 engine functions on 24 volts. The distributor is similar in design to the 1949–1953 Ford flathead V-8 distributor except that it is a dual-point model. It is hard to believe the French military equipped its flatheads with a dual-point distributor right up to the 1990s, while every other engine manufacturer had converted to electronic ignitions years before. That is a negative point for the French flathead.

The ignition system is a very key part of a blown street-performance Ford flathead V-8 engine. It is mandatory to install a modern electronic distributor to prevent ignition failure, particularly when the engine is experiencing boost. If the ignition were to fail while the engine was experiencing boost, the result would be a disaster of major proportions to the engine and the owner's wallet.

The MSD 8573 electronic distributor with vacuum advance is a good distributor for 1949–1953 Ford flathead V-8 engines. This distributor is referred to as a "ready-to-run" model because it has a built-in inductive ignition module. The MSD 8574 Pro-Billet distributor with full centrifugal (mechanical) advance is also available for 1949–1953 Ford flathead V-8 engines, although this model must be used with the MSD 6A or 6AL control box. These distributors are machined from 6061 T-6 aluminum to within 0.001 inch. There is an upper sealed ball bearing unit, the shaft is polished and Tuftrided, and they will operate through 10,000 rpm. The mechanical advance in the MSD 8574 distributor is adjusted by changing the color-coded springs, which are heavy silver, light silver, or light blue. The MSD 8573 or 8574 distributors do not have an extended shaft, meaning a 1949–1950 Ford 6059 cast-iron timing gear cover should be used with them. The current retail price for either of these distributors is $340. The distributors come equipped with a distributor gear; however, the shaft must be drilled to install the gear. We installed an MSD 8574 Pro-Billet distributor in the French flathead V-8 engine.

The Mallory (Mr. Gasket) 3727501 Unilite distributor is triggered by a self-contained, photo-optic infrared LED system, and the Mallory 5027501 is a magnetic breakerless distributor. These are excellent distributors with full centrifugal (mechanical) advance, they are machined from 6061 T-6 aluminum, there is an upper sealed ball bearing unit, the shaft is heat-treated and centerless ground, and they will operate through 10,000 rpm. The mechanical advance is adjusted with degree keys and springs. Each of these two distributors has an extended shaft, meaning a 1951–1953 Ford C8RT-6059-B/C0BA-6059-A or C8RT-6059/OBA-6059-B cast-aluminum timing gear cover should be used with them. The current retail price for either of the Mallory distributors is $320. The Mallory 3727501 and 5027501 distributors are not sold with a distributor gear.

The 1949–1950 Ford 12390-C distributor gear was secured on the distributor shaft with the Ford 72837-S rivet. The 1951–1953 Ford 12390 distributor gear was secured on the distributor shaft with the Ford 61465-S pin. Other than that, the distributor gears are the same. The distributor gear must be installed at the correct position on the distributor shaft in order to correctly engage the gear on the nose of the camshaft. The distributor has a flange on the upper housing where the 1949–1953 Ford 12270 distributor clamp attaches to secure the distributor in position. The distance from the bottom of that flange to the bottom of the distributor gear must be 8.00 inches (+ or – 0.020 inches). The end play on the distributor shaft

An MSD 8574 Pro-Billet electronic distributor for 1949–1953 Ford flathead V-8 engines; a 1949–1950 Ford 6059 chrome cast-iron timing gear cover.

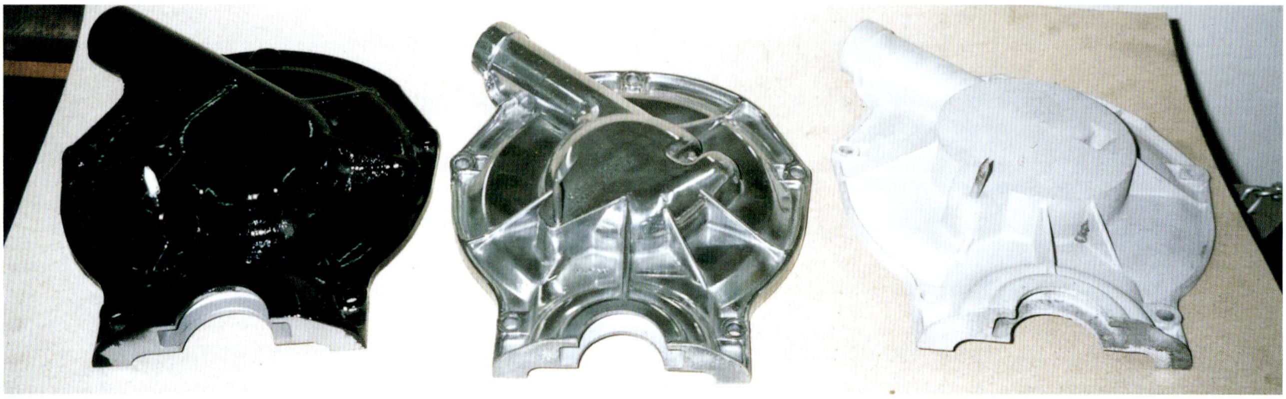

On the left, a 1949–1950 Ford 6059 cast-iron timing gear cover. In the center, a 1951–1953 Ford C8RT-6059-B/COBA-6059-A cast-aluminum timing gear cover. On the right, a 1951–1953 Ford C8RT-6059/OBA-6059-B cast-aluminum timing gear cover.

should be 0.002 to 0.005 inches after the distributor gear is installed; shim accordingly. Use a polished stainless-steel 5/16-inch N.C. by 1-inch-long hex head bolt and lock washer with the 1949–1953 Ford 12270 distributor clamp. The Speedway Motors 910-12701 distributor gear is a replacement gear for the 1949–1953 Ford flathead V-8 distributors; the current retail price for this part is $15.

The 1949–1950 Ford 6059 cast-iron timing gear cover was designed for use with a distributor without an extended shaft. The internal length of the neck on this timing gear cover is long in order to support the body of the distributor. The MSD distributor should be used with this timing gear cover. The 1951–1953 Ford C8RT-6059-B/COBA-6059-A cast-aluminum timing gear cover and the 1951–1953 Ford C8RT-6059/OBA-6059-B cast-aluminum timing gear cover have a distributor shaft boss on the inside used to support the end of the distributor shaft, which permits the internal length of the neck in these timing gear covers to be shorter. The Mallory distributors should be used with cast-aluminum timing gear covers.

If an aftermarket billet-aluminum distributor is being used in a Ford flathead V-8 engine equipped with an aluminum timing gear cover, the steps on the shaft housing should be measured prior to installing the distributor. The maximum diameter of the first step should be 1.058 inches, the second step should be 0.930 inch, and the step near the distributor drive gear should be 0.900 inch. I recently encountered an aftermarket distributor where the diameter of the second step was 0.950 inch. That distributor could not be installed until 0.020 inch was removed in a lathe from the second step.

Coil and Ignition Control

Refer to Chapter 7 for information pertaining to an ignition coil and ignition control boxes. If an aftermarket distributor will be installed in a flathead engine equipped with a

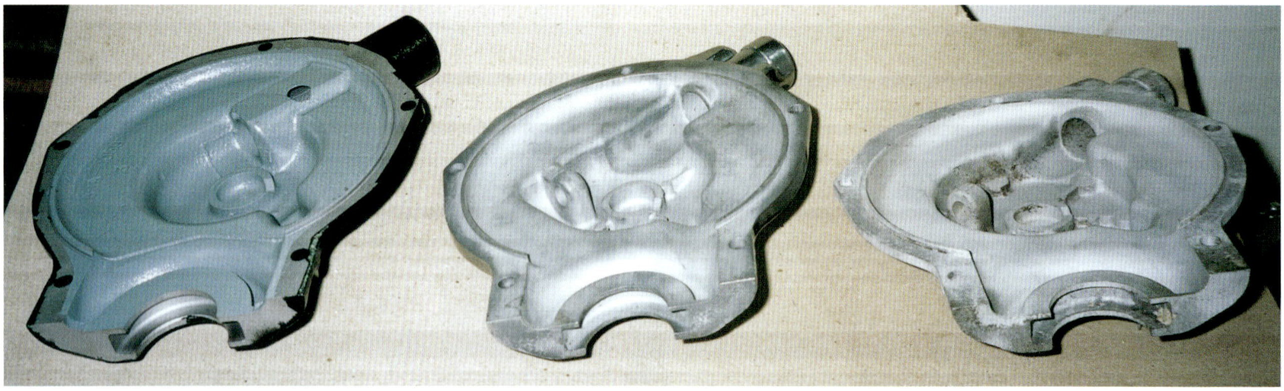

The timing gear cover on the left is a 1949–1950 6059 cast-iron model. The timing gear cover in the center is a 1951–1953 Ford C8RT-6059-B/COBA-6059-A cast-aluminum model. The timing gear cover on the right is a 1951–1953 Ford C8RT-6059/OBA-6059-B cast-aluminum model. Each aluminum timing gear cover has a distributor shaft support boss.

1949–1953 timing gear cover and 1949–1953 cylinder heads, an offset coil bracket will be required. Aftermarket distributors are much bulkier than the stock 1949–1953 Ford distributors. If the ignition coil is mounted in the stock position on the right (passenger's) side cylinder head, the top of the ignition coil will make contact with the aftermarket distributor cap. Luke's Custom Machine & Design fabricates a polished-aluminum offset ignition coil bracket to overcome this problem.

Spark Plug Wires and Plugs

Refer to Section I

Alternator

The French flathead V-8 was equipped with a 50-amp generator similar to the 1949–1953 Ford flathead V-8 generator, except that it was designed for a 24-volt electrical system. The French generator was manufactured in France (Paris–Rhone) and the part number G13R40 has been observed on one of them.

We installed a new (Nippon) Denso 16241-64011 alternator on the blown French flathead V-8 engine. The alternator is mounted with the stud in the front of the intake manifold used for the stock flathead generator.

Here is an original Mallory 7C Model 275A dual-point distributor for a 1949–1953 Ford flathead V-8 engine. The diameter of the first step is 1.058 inches, the second step is 0.930 inch, and the third step is 0.900 inch. This distributor is equipped with an advance unit for use with later-model Holley 94 carburetors.

Starting Motor

The French flathead V-8 was equipped with a 24-volt starting motor. The flywheel used with the starting motor is different from the 1946–1953 Ford flathead V-8 flywheels. Powermaster has recently started to market its new ministarter for the French flathead V-8 engine. The new model is the Powermaster 9508 XS torque starting motor, which weighs a mere 8 pounds and is said to create more than double the peak horsepower torque of the French flathead V-8 model.

We installed a 1946–1948 Ford flathead V-8 flywheel in order to be compatible with the installed Model 59 oil pan and the starting motor positioning. We also installed a Powermaster 9507 XS torque ministarter.

Part B

- New MSD 8574 Pro-Billet electronic distributor, 1949–1953 Ford, (CNC machined 6061 T-6 aluminum) with full centrifugal (mechanical) advance (25 degrees advance at 2,500 rpm) installed with 1949–1953 Ford 12270 distributor hold-down clamp and new polished stainless-steel head hex bolt and lock washer using antiseize compound; new MSD 6420 Model 6AL multiple-spark discharge ignition control with rpm-limiter modules and new MSD 8200 Blaster 2 chrome ignition coil (45,000 volts) installed with 1949–1953 Ford 12043 ignition coil bracket and Luke's Custom Machine & Design new polished-aluminum offset ignition coil bracket using new polished stainless-steel hex head bolts and lock washers; Ford ignition coil bracket painted with PPG 7400 red high-gloss polyurethane; **$615.52**
- New Pertronix Performance Products 808290 Flame Thrower high-performance 8-millimeter spark plug wires (black 90-degree boots) installed with new Spectre 4245 chrome professional wire separators; new NGK B-6L spark plugs installed with 0.050-inch gap and torqued to 15 ft-lb using antiseize compound; **$76.70**
- New (Nippon) Denso 16241-64011 polished alternator (35 amps, internally regulated) installed with a Luke's Custom Machine & Design new polished-aluminum alternator pulley and new polished stainless-steel alternator brace using a new polished stainless-steel AN flat washer, lock washer, and nut with antiseize compound; **$310.87**
- New Powermaster 9507 XS torque starting motor installed; **$189.95**

Ignition System Total: $1,193.04

CHAPTER 18
COOLING SYSTEM AND MISCELLANEOUS

Part A
Water Pumps

The water pumps supplied with the French flathead V-8 engine have a horizontal motor mount flange cast into them similar to the 1949–1953 Ford Truck models. The left (driver's) side water pump motor mount flange also has a horizontal tube cast into the end. The water pump pulleys are designed for a single V-belt; however, the left water pump pulley is smaller in diameter than the right water pump pulley. The left and right water pump pulleys for the French flathead V-8 engine are not the same distance from the front face of the engine block; they are staggered. This is a situation similar to 1949–1953 Ford flathead V-8 passenger car water pumps.

There is a water pump bypass hole on both sides of the front of the French flathead V-8 engine block that exits at the top front of the deck surface on both sides of the engine block. This is similar to 1949–1953 Ford flathead V-8 engine blocks. The water pump bypass holes at the front of the engine block and the top of the deck surfaces had previously been plugged. The engine block had been ground in the area above the top of the water pumps in order to follow the contour of 1937–1948 Ford flathead V-8 water pumps. These modifications to the French flathead V-8 engine block prohibited the future use of 1949–1953 Ford flathead V-8 water pumps. Refer to Chapter 8 for more information.

Because there were no water pump lubrication holes in the front of the French flathead V-8 engine block, we installed 1937–1941 Ford Truck 79-8505 and 79-8506 water pumps. These water pumps were cleaned, glass beaded, sanded and detailed, painted, and rebuilt. In addition,

These 1937–1941 Ford truck 79-8505 and 79-8506 water pumps are equipped with aluminum pulleys fabricated by Luke's Custom Machine & Design.

Luke's Custom Machine & Design fabricated the polished-aluminum single-groove water pump pulleys.

Labor

The cost of the labor for checking clearances, gapping piston rings, degreeing the camshaft, painting and detailing, trial assembly of motor, final assembly of motor, blower setup, and the initial startup of the engine are not included in Part B of each chapter in this section of the book. I shall now include that cost here, at the end of the chapter.

Part B

- 1937–1941 Ford Truck 79-8505 and 79-8506 water pumps rebuilt with NOS Ford 79-8591 repair kits and installed with new polished stainless-steel hex head bolts and lock washers using Loctite and torqued to 25 ft-lb; water pump gaskets installed using silicone sealant; Luke's Custom Machine & Design new polished-aluminum water pump pulleys installed; new Gates 11A1255 XL crankshaft/water pumps/alternator V-belt (11/32 inch by 49 1/2 inches long) installed; water pumps painted with PPG 74000 red high-gloss polyurethane; **$505.37**
- Labor for checking clearances, gapping piston rings, degreeing the camshaft, painting and detailing, trial assembly of motor, final assembly of motor, blower setup, and the initial startup of the engine; **$1,733.33**

Cooling System and Miscellaneous Total: $2,238.70

Note: The estimated output of this engine is 281 horsepower at 5,000 rpm and 323 ft-lb torque at 3,500 rpm (See Appendix A: Dyno Print-Outs).

Blown French Flathead Street Engine Total:

$4,216.39 engine block total
$1,166.46 crankshaft total
$1,713.61 connecting rods and pistons total
$822.36 lubrication system total
$1,931.00 camshaft and cylinder heads total
$4,827.97 intake system total
$1,193.04 ignition system total
$2,238.70 cooling system and miscellaneous total
$18,109.53 (U.S.)

Opposite: *The top of the engine block was ground to follow the shape of the 1937–1948 Ford flathead V-8 water pumps. A 1949–1953 Ford truck water pump is bolted in place. The water pump bypass hole is exposed, preventing the future use of this type of water pump on this particular engine block.*

CHAPTER 19
MISSING PARTS

Exhaust Manifolds

The exhaust manifold bolt holes in the French flathead V-8 engine block are the same as the exhaust manifold bolt holes in the 1946–1953 Ford flathead V-8 engine block. The two front holes on each side of the engine block have 7/16-inch N.C. threads, and the rest of the holes have 3/8-inch N.C. threads. The French flathead engine block was fitted with studs for the N.C. threads, but they also have 10-millimeter by 1.25-pitch threads on the exhaust manifold end. The studs should be retained because the engine block bolt holes are open to the water jackets. If bolts were used for the exhaust manifolds, engine vibration might cause them to loosen, resulting in a water leak. The French flathead engine is equipped with cast-iron exhaust manifolds. The left exhaust manifold has a casting number of 710064, and the right exhaust manifold has a casting number of 710063, with both exhaust manifolds exiting at the rear of the engine.

The exhaust manifolds have not been included in the description of the engine in this section of the book because the type of exhaust header will depend on the construction of the frame that the engine will be installed in. The Speedway Motors 910-13605 exhaust manifolds are the original Fenton 07101LH and 07101RH cast-iron models and are available at the current retail price of $290. The Speedway Motors 910-13601 exhaust headers are a center dump type and are available for $150; the 910-13601H exhaust headers with metallic-ceramic coating are available for $220. The Speedway Motors 910-13606 economy-model exhaust headers are available without collector flanges for $120. Use high-temperature silicone sealant when installing the exhaust manifold gaskets, and torque the exhaust manifold stud nuts to 25 ft-lb using Loctite.

A blown Ford flathead V-8 engine requires a good exhaust system, and proper tube exhaust headers are ideal for this type of engine. Tatom Custom Engines (see Appendix B: Resources) sells nice 14-gauge-steel equal-length exhaust headers with 3/8-inch-thick flanges. These headers are available for most early Ford hot rod chassis.

This Ford flathead V-8 engine is equipped with Fenton metallic-ceramic-coated cast-iron exhaust headers, good for a normally aspirated engine. The Offenhauser 5272-A transmission adapter is visible.

Luke Balogh fabricated this equal-length exhaust header for his 1950 Meteor. The end tubes are 1 1/2 inches in diameter, and the center tube is 1 7/8 inches in diameter. This type of exhaust header is a good choice for a blown flathead.

Port match the exhaust headers to the exact shape of the exhaust ports in the engine block prior to having the exhaust headers metallic-ceramic coated.

Flywheel and Flexplate

I described some popular transmissions that can be adapted to a Ford flathead V-8 engine block in my first flathead book and will repeat only one example here. The French flathead V-8 engine described in this section of the book will have a 1985–1993 Chevrolet S-10 pickup manual five-speed transmission installed with a 1946–1948 Ford 6375 flywheel, a 1949–1953 Ford 7563 long-style 11-inch-diameter pressure plate, a Tredex Automotive TD4212 clutch disc (for a 1985–1991 Chevrolet Astro Van) with 14 splines of 11-inch diameter, and a bronze pilot bearing that is 1.575 inches o.d. by 0.595 inch i.d. You will also need an Offenhauser 5272 transmission adapter. Refer to Chapters 9 and 29 for more information about adapting transmissions to the Ford flathead V-8 engine.

These parts are required to adapt a 1985–1993 Chevrolet S-10 pickup manual five-speed transmission to a 1949–1953 Ford flathead V-8 engine: a 1949–1953 Ford 6375 flywheel with a 1.575-inch i.d and 0.595-inch i.d. bronze pilot bearing, a Tredex Automotive TD4212 clutch disc with 14 splines and 11-inch diameter, and a 1949–1953 Ford 7563 long-style 11-inch-diameter pressure plate.

Fan

Refer to Section I

Power Steering

Ford Motor Company offered a power steering option for the 1953 Ford and Mercury passenger cars. The parts that make up this kit are next to impossible to locate today and, even if they are found, most likely they will be worn out. Luke Balogh has fabricated a power steering package for his 1950 Meteor (1950 Canadian Ford) coupe that consists of a GM 7808286 power steering pump (1965–1973 Corvette), a GM 1493256 polished-aluminum power steering pump bracket (1970 Cadillac), and a custom-polished stainless-steel cylinder head mounting bracket for the power steering pump. The neck of the power steering pump was cut off, and a finned reservoir was adapted to the power steering pump. A GM 5691955 steering gear valve (1963–1973 Corvette) will be fitted to the steering relay rod.

Luke's Custom Machine & Design assembled the completed power steering pump and brackets for a Ford flathead V-8 engine.

CHAPTER 20
ASSEMBLY

Note: Be prepared to spend a lot of additional time assembling this engine; if an engine building shop is doing the work, be prepared to spend a lot of money. Many of the Ford flathead V-8 and aftermarket parts must be custom fitted to the French flathead V-8 engine block. It costs an estimated $2,000 more to use a French flathead V-8 engine block than a Ford flathead V-8 engine block as the basis for a street-performance engine. Also refer to Sections I and III.

Engine Block

Clean the camshaft bearings with solvent using a clean, lint-free cloth. Install the camshaft bearings dry in the engine block using a camshaft bearing installation tool (this is a must), taking special care to line up the oil holes in the camshaft bearings with the oil holes in the engine block. Do not lubricate the camshaft bearing bosses.

Install the oil pump idler gear cover gasket using silicone sealant. Lubricate the oil pump idler gear bushing and oil pump idler gear shaft with motor oil, and then install the assembly in the rear of the engine block. The smaller side of the oil pump idler gear faces outward. Install the oil pump idler gear cover bolts and lock washers using Loctite, and torque them to 15 ft-lb.

Install a new stainless-steel 1/8-inch NPT plug in the oil line at the face of the engine block (behind the camshaft gear) using Permatex aviation form-a-gasket. Install new stainless-steel 1/4-inch NPT plugs in the water jacket drain holes (on both sides of the front of the engine block near the oil pan rails) using Permatex aviation form-a-gasket. Install the aluminum block-off plug for the fuel pump pushrod bushing using Permatex aviation form-a-gasket.

Install the camshaft gear on the camshaft using Loctite on the bolts, and torque the bolts to 20 ft-lb. Bend the camshaft locking plate tabs over the heads of the camshaft bolts.

Lubricate the camshaft bearings and the camshaft bearing journals with Clevite 77 Bearing Guard, and generously apply the camshaft assembly lubricant (provided by the camshaft manufacturer) to the lobes of the camshaft. Gently slide the camshaft into the engine block, taking care not to nick the camshaft bearings.

The French flathead engine block has been thoroughly cleaned, professionally painted, and is ready for assembly. Notice that the exhaust manifold studs have been installed.

Crankshaft

Install the rear main bearing seal retainer in the engine block using silicone sealant, and gently tap it into position with a rubber mallet. Install the rear upper cord-material oil seal in the retainer by rolling it into place with a large socket, or use the head of a large ball-peen hammer and tap the head of the hammer with another hammer. Repeat this method when installing the rear lower cord-material oil seal in the rear main bearing cap.

Clean the main bearings with solvent using a clean, lint-free cloth. Line up the locking tang (tab) on the main bearings with the notch in the main bearing saddles, and install the main bearings in the engine block dry. Do not lubricate the main bearing saddles in the engine block. Use Clevite 77 Bearing Guard to lubricate the face of the main bearings after they are installed in the engine block. Line up the locking tang (tab) on the main bearings with the notch in the main bearing caps, and install the main bearings in the main bearing caps dry. Do not lubricate the main bearing caps. Use Clevite 77 Bearing Guard to lubricate the face of the main bearings after they have been installed in the main bearing caps.

Install the stainless-steel connecting rod journal oil line plugs in the crankshaft using Loctite. Place the crankshaft gear Woodruff key in the crankshaft using antiseize compound, and gently tap the Woodruff key into position with a hammer. Apply antiseize compound to the snout of the crankshaft, and slide the crankshaft gear onto the snout as far as it will go by hand. The side with the dot or line on the crankshaft gear faces outward. Line up the crankshaft gear with the crankshaft gear Woodruff key, and gently tap the crankshaft gear into position using a piece of steel pipe a little larger in diameter than the diameter of the crankshaft snout.

Gently lower the crankshaft into the main bearing saddles, taking special care not to nick the crankshaft bearings. Make certain to line up the timing marks on the crankshaft and camshaft gears. The crankshaft gear has a dot or a line, and the camshaft gear has a dot or a line. These marks should be vertically in line with each other. Place the main bearing caps in their correct order in the engine block, with the tang (tab) in the engine block and the tang (tab) in the main bearing caps on the same side. Lubricate the main bearing cap bolts with molykote, and torque the main bearing cap bolts to 105 ft-lb. The crankshaft should now turn freely in the engine block. Slide the oil slinger onto the crankshaft snout.

The crankshaft has been installed and the main bearing cap bolts properly torqued. Notice the Doug King Enterprises number 1 and 2 main bearing caps and the exhaust baffle in the center exhaust port.

Connecting Rods and Pistons

Remove the wrist pins from the pistons. Coat the wrist pin bushings in the connecting rods and the wrist pins with motor oil. The bearing notch in the connecting rods faces outward from the crankshaft. Assemble the connecting rods and pistons, and install the Spiro-Lox retainers.

Install the piston rings on the pistons using a piston ring expander, and rotate each of the piston rings on the pistons so that they are within the manufacturer's recommended arc, also known as "preferred ring gap location."

The Ross Racing Pistons have been assembled with SCAT Enterprises H-beam connecting rods. The piston rings were installed using the piston ring expander in the foreground. The assemblies are now ready for installation.

Clean the connecting rod bearings with solvent using a clean, lint-free cloth. Install the connecting rod bearings dry in the connecting rods and the connecting rod caps. Line up the locking tang (tab) on the connecting rod bearing with the notch in the connecting rod and the connecting rod cap. Apply Clevite 77 Bearing Guard to the face of the connecting rod bearings.

Apply motor oil to the inside of the piston ring compressor, and place the piston ring compressor around the piston. Tighten the piston ring compressor, insert the connecting rod into the cylinder (with the connecting rod bearing locking tang facing outward from the crankshaft), and gently tap the piston down into the cylinder using a rubber mallet or a piece of wood. Guide the connecting rod over the connecting rod journal as it is being pushed downward. After the connecting rod is on the connecting rod journal, install the connecting rod cap with the locking tang (tab) on the same side as the locking tang (tab) in the connecting rod. Install the connecting rod bolts using molykote, and torque the connecting rod bolts to 42 ft-lb, allowing for 0.0061-inch stretch. Repeat this procedure for the rest of the connecting rod and piston assemblies.

Install the crankshaft bolt in the crankshaft and turn the crankshaft. The rotating assembly should turn over smoothly.

Use a feeler gauge to check the side clearance for each pair of connecting rods.

The connecting rod and piston assemblies have been installed, and the connecting rod bolts have been torqued to 42 ft-lb. Notice the cork rear oil pan seal on the rear main bearing cap.

Lubrication System

Install a stainless-steel 1/4-inch NPT plug inside the vertical drain-back oil line hole in the engine block using pipe thread sealant. Install the oil line fittings for the pressure-out oil line and the drain-back oil line using pipe thread sealant.

Lubricate the oil pump drive gear with motor oil, and install the oil pump assembly in the engine block. Install the oil pump bolt and lock washer using Loctite, and torque the bolt to 15 ft-lb.

Slide the crankshaft sleeve onto the crankshaft, and gently tap it into position using a piece of steel pipe and a hammer. Slide the one-piece oil seal over the crankshaft sleeve. Install the timing gear cover gasket using silicone sealant, and apply silicone sealant to the inside of the timing gear cover where the oil seal is located. Install the timing gear cover and the timing gear cover bolts and lock washers using Loctite, and torque them to 15 ft-lb.

A flywheel or flexplate will have to be installed in order to start the engine. Install the flywheel or flexplate prior to installing the oil pan. Install the flywheel or flexplate bolts using Loctite, and torque the bolts to 80 ft-lb.

Install the cork material gasket around the end of the rear main bearing cap using silicone sealant. Place some putty on the oil pump pickup screen, and position the oil pan on the engine block without the oil pan gaskets. Push down firmly on the oil pan, and then remove the oil pan. Measure the compressed thickness of the putty. The

The oil pump and pickup have been installed, and the clearance with the oil pan has been checked.

oil-pan-to-oil-pump-pickup-screen clearance should be 3/4 to 1 inch. If the clearance is correct, carry on. If the clearance is not correct, adjust the oil pump pickup screen arm until the correct clearance is obtained. Install the oil pan gaskets using silicone sealant. Install the oil pan and the oil pan bolts, lock washers, and AN flat washers using Loctite, and torque the oil pan bolts to 15 ft-lb. Screw the dipstick tube into the oil pan using silicone sealant, and install the dip stick.

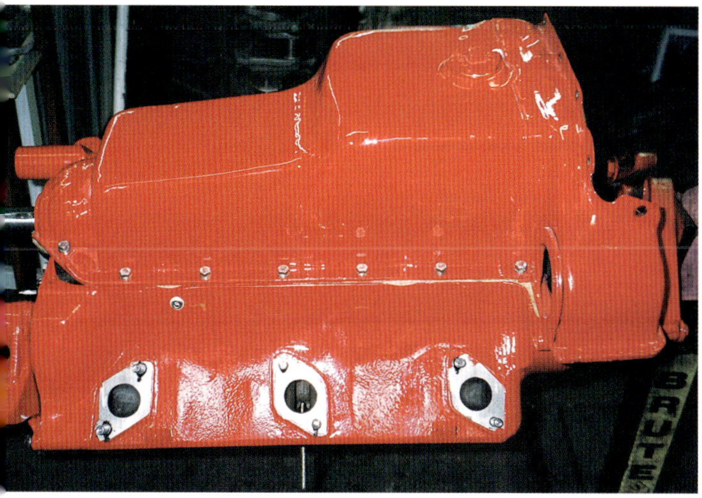

Once the oil pan has been installed, flip the engine over, and get started on the top side.

Install the oil line from the pressure-out fitting in the engine block to the inlet side of the oil filter adapter using pipe thread sealant. Install the oil line from the drain-back line to the outlet side of the oil filter adapter using pipe thread sealant. Fill the oil filter with motor oil, and install it on the oil filter adapter.

Pour the required amount of oil in the engine, and install a temporary oil-pressure gauge.

Camshaft and Cylinder Heads

Lubricate the body of the intake and exhaust valve lifter (tappet) for the number 1 cylinder with motor oil and the face of each valve lifter with the camshaft assembly lubricant supplied by the camshaft manufacturer. Install the two valve lifters and degree the camshaft. If the results of degreeing the camshaft are in accordance with the camshaft manufacturer's specifications (timing tag), continue on with the installation of the rest of the valve lifters. If the results are not the same, find out why before proceeding.

Place the dual valve springs with the dual valve spring retainer and Ford valve spring retainer and sleeve in position in the valve lifter gallery and insert the valve guide,

All the valves have been installed and adjusted.

followed by the crow's foot valve guide retainer. Insert the valve through the valve guide and the valve spring retainers and sleeve. Use a valve guide removal tool to lift the valve spring retainers and sleeve high enough that the valve stem locks can be installed. It is much easier to do this job with two people, although even then it is a painfully slow process. Repeat this procedure with the rest of the valve assemblies. Insert a 3/16-inch drill bit (use the chuck end) or punch into the hole drilled in the valve lifter boss, and adjust the valve according to the camshaft manufacturer's specifications. Repeat this process for all the valves. Rotate the crankshaft until the exhaust valve for that cylinder starts to open, and adjust the intake valve for that cylinder. Rotate the crankshaft until the intake valve for that

The fuel pump pushrod bushing plug, custom crankcase breather tube, Ford 6524 valve lifter gallery oil-return hole baffles, and Iskenderian Racing Cams 4005 dual valve springs are all visible in this picture.

cylinder starts to close, and adjust the exhaust valve for that cylinder. Adjust all the valves for all the cylinders in the same manner. Install the valve lifter gallery crankcase breather tube by gently tapping it in position with a rubber hammer. Install the two valve lifter gallery oil return hole baffles by snapping them into position.

Put some putty on any intake valve and exhaust valve for the same cylinder and then place the cylinder head on the engine block and lightly tighten the cylinder head down using two cylinder head bolts (one at each end of the cylinder head). Do not use the cylinder head gasket. Turn the crankshaft through 360 degrees at least twice. Remove the cylinder head and measure the thickness of the putty. The absolute minimum valve-to-cylinder-head clearance should be 0.040 inch.

Spray the copper-sandwiched cylinder head gaskets with Permatex copper spray-a-gasket. Install the cylinder head gaskets and cylinder heads with the cylinder head bolts and washers using Permatex aviation form-a-gasket. Torque the cylinder head bolts to 40 ft-lb (for aluminum cylinder heads) using the factory-approved torque pattern. Do not install the chrome acorn nuts until the engine has been initially started and the cylinder heads have been retorqued.

Install the thermostats in the cylinder heads. Install the water outlet elbow gaskets using silicone sealant. Install the water outlet elbows using antiseize compound on the bolts, and torque the water outlet elbow bolts to 15 ft-lb. Install the temperature-sending unit in the left (driver's) side cylinder head using pipe thread sealant. Install stainless-steel 3/8-inch NPT plugs in any unused cylinder head openings.

The valves are covered with putty to check the valve-to-cylinder-head clearance. The clearance was 0.200 inch for the intake valve and 0.210 inch for the exhaust valve.

The cylinder head bolt torque sequence recommended by Ford Motor Company applies to all 24-stud flathead engines.

The Edelbrock 1115 cylinder heads have been installed and torqued, and the modified water outlet elbows have also been installed.

Intake System

Trace the intake manifold outline on the intake manifold gasket, and trim the intake manifold gasket accordingly. Install the intake manifold gasket using silicone sealant. Install the intake manifold and torque the intake manifold bolts to 25 ft-lb using Loctite. Torque the intake manifold bolts using the factory-approved torque pattern (refer to Chapter 6). Install the chrome acorn nut covers on the intake manifold bolts. Install the fuel pump/crankcase breather housing using silicone sealant on the gasket, and torque the acorn nuts to 15 ft-lb using antiseize compound.

Install the blower assembly on the intake manifold using Permatex aviation form-a-gasket on the intake-manifold-to-blower gasket. Torque the blower-to-intake-manifold stud nuts to 25 ft-lb using antiseize compound. Install the blower snout assembly on the front cover of the blower using Permatex aviation form-a-gasket on the blower snout gasket. Install the blower snout bolts using antiseize compound, and torque the blower snout bolts to 20 ft-lb. Install the blower snout pulley using antiseize compound on the blower snout pulley bolts and torque them to 25 ft-lb. Pour enough gear oil into the blower front cover so that it is about halfway up the oil level sight glass.

Install the carburetor-to-blower adapter using Permatex aviation form-a-gasket on the gasket. Install the carburetor adapter bolts using antiseize compound, and torque them to 20 ft-lb. Install the carburetor studs using antiseize compound. Install the carburetor with the carburetor gasket, and tighten the carburetor stud nuts snugly; do not overtighten them. Install the dual inlet fuel line in the carburetor. Install the fuel-pressure regulator in the dual inlet fuel line. Install the fuel-pressure gauge in the fuel-pressure regulator.

Install the blower belt idler pulley on the blower belt idler pulley bracket using antiseize compound on the bolt. Install the blower belt idler pulley bracket on the front of the intake manifold using antiseize compound on the stud.

Ignition System

Install the crankshaft pulley Woodruff key in the crankshaft slot using antiseize compound and gently tap the Woodruff key into position with a hammer. Apply antiseize compound to the snout of the crankshaft, and install the one-piece crankshaft hub and pulley. The pulley should have a snug fit but it should not require the use of a harmonic balancer installation tool. Gently tap the crankshaft hub and pulley onto the crankshaft snout using a rubber mallet or a piece of wood and a hammer. Install the crankshaft pulley bolt and washer using Loctite, and torque the crankshaft bolt to 50 ft-lb. Install the crankshaft blower pulley using antiseize compound on the crankshaft blower pulley bolts, and torque them to 25 ft-lb.

Rotate the crankshaft until the engine is set with the number 1 piston at 10 degrees before top dead center (TDC), and insert the distributor in the timing gear cover with the distributor rotor at the number 1 spark plug lead position. Install the distributor clamp and bolt using antiseize compound and tighten the bolt snugly.

Gap the spark plugs at 0.050 inch and install them using antiseize compound. Torque the spark plugs to 15 ft-lb. Lay out the spark plug wires, and cut them to length for each cylinder. The cylinders on the right (passenger's) side are numbered 1, 2, 3, and 4 from the front of the engine block. The cylinders on the left (driver's) side are numbered 5, 6, 7, and 8 from the front of the engine block. The firing order is: 1, 5, 4, 8, 6, 3, 7, and 2. Install the spark plug wires on the spark plugs and in the distributor cap. Install the offset ignition coil bracket using antiseize compound on the bolts, and torque them to 15 ft-lb. Install the ignition coil bracket on the offset ignition coil bracket using antiseize compound on the bolts, and torque

The blower has been mounted on the intake manifold, the carburetor adapter bolted to the blower, the carburetor bolted to the carburetor adapter, and the air cleaner has been installed.

Luke's Custom Machine & Design fabricated this polished stainless-steel alternator adjusting brace.

them to 15 ft-lb. Install the ignition coil in the ignition coil bracket. Connect the coil wire to the distributor.

Install the alternator on the generator mounting stud, and install the alternator adjusting brace using antiseize compound on the bolt.

Install the starting motor using antiseize compound on the bolts.

The distributor with spark plug wires, the water pumps, the alternator, and the V-belt have been installed.

The blower belt has been installed, allowing for 3/4-inch slack.

The Auto-Meter 4401 Pro Comp Ultra-Lite vacuum/boost gauge and the 4411 fuel-pressure gauge are mounted at the rear of the blower case.

Cooling System and Miscellaneous

Apply silicone sealant to the water pump gaskets and the bolts. Install the water pumps, and torque the water pump bolts to 25 ft-lb.

Install the crankshaft pulley/water pumps/generator V-belt, allowing for 3/4 to 1 inch slack.

Install the blower belt and adjust it, allowing for approximately 3/4 inch slack

Install the boost gauge mounting bracket with the boost gauge, and connect the boost gauge hose from the intake manifold fitting to the boost gauge.

Install the PCV valve in the intake manifold and the hose line fittings in the oil separator using pipe thread sealant. Bolt the oil separator to the right (passenger's) side of the back of the blower case. Connect the hoses from the PCV valve to the oil separator and the oil separator to

This is a rear view of the completed French flathead V-8 engine. The oil separator and crankcase breather/oil filler housing look compatible with this engine.

This is a right-hand view of the completed French flathead V-8 engine.

the PCV fitting at the rear of the Holley carburetor. Tighten the hose clamps.

Bolt on the headers and mufflers and a radiator with some water in it, wire the ignition system, and hook up a battery with a starter switch. Start the engine, and adjust the timing and the carburetor. Allow the engine to run at approximately 2,000 rpm for 30 minutes in order to break in the camshaft. After the engine is at full operating temperature, shut it off, and allow the engine to cool. Retorque the cylinder heads using the factory-approved torque pattern, and install the chrome acorn nut covers on the cylinder head bolts.

This engine has a nice torque curve from 2,000 to 4,000 rpm, indicating the correct combination of engine parts was installed.

SECTION III: BLOWN ARDUN STREET-PERFORMANCE ENGINE

CHAPTER 21
ENGINE BLOCK

In 1947, Zora Arkus-Duntov (1909–1996) and his brother, Yura Arkus-Duntov (1917–1980), completed the design of the Ardun overhead valve (OHV) hemispherical combustion chamber conversion kit for the Ford flathead V-8 engine. The brothers were living in New York at the time, and their company was named Ardun Mechanical Corporation. The name "Ardun" is derived from letters within their last names. Zora Duntov went on to become one of the most influential forces in the development and refinement of the Corvette. The famous Duntov camshaft for the small-block Chevrolet engine is named after him. The introduction of the Ardun cylinder heads revolutionized the American automotive engine industry; these were the forerunners to the early Chrysler Hemi cylinder heads.

The Ardun cylinder heads were cast in England using Alcoa 355 T-6 heat-treated aluminum alloy. It is rumored only 200 sets of these cylinder heads were produced, although no one seems to know for certain. The Allard Motor Company sold Ardun V-8 conversion kits all over Europe and provided brochures in English, French, and German. It is unknown how many sets of the original Ardun cylinder heads remain in working condition today. The literature provided with the Ardun V-8 conversion kit states that the installation requires no special tools and can be carried out in six to eight hours. This is slightly misleading advertising. Many of the components must be custom fitted by a competent machine shop, which is one of the reasons that the Ardun V-8 conversion kit was not an immediate marketing success. Even today, there are very few people in the world that can properly build an Ardun V-8 engine.

Don Ferguson of Ardun Enterprises (see Appendix B: Resources) produces an improved and premium-quality Ardun V-8 conversion kit. These new cylinder heads are manufactured using high-grade Alcoa 356 T-6 heat-treated aluminum and they are CNC machined. The complete kit is available for the current retail price of $12,990.

The owner of the Ardun engine described in my first flathead book had another set of Ardun cylinder heads that were part of a complete engine. This was the famous Ardun/Mercury V-8 drag race engine built by T. "Sooey" Suenega of Hawaii Motor Rebuilders in 1955. The engine was originally installed in a white 1932 Ford five-window coupe that was featured in the October 1957 and August 1960 issues of *Hot Rod* magazine. Hawaii Motor Rebuilders was the only company in Hawaii in 1956 with an engine-balancing machine. Sooey maintained this was one of the reasons his car consistently ran 120 miles per hour and won most of the races entered.

The National Hot Rod Association (NHRA) 1956 National Championship Drag Races were held in Kansas City, Missouri, from August 31 to September 1. The Hawaii

The Ardun cylinder heads from the Hawaii Motor Rebuilders drag race engine are shown in the trunk of the owner's beautiful Ardun-powered 1932 Ford roadster.

Motor Rebuilders 1932 Ford won the 1956 Nationals B/Altered class. The engine was equipped with an Iskenderian Racing Cams Crossflow camshaft and Hilborn fuel injection.

Suenega sold the Hawaii Motor Rebuilders 1932 Ford five-window coupe in 1958 to the Dowsett Racing Team. The complete Ardun/Mercury V-8 engine from that car was eventually sold to the party that the current owner purchased the engine from in 1976 for the unbelievable price of $800. Now, that is an investment!

The Ardun engine had been in storage since 1976. Then, in July 2003, the owner decided to have a blown Ardun street-performance engine built. Work commenced with the delivery of the cylinder heads to Luke's Custom Machine & Design. These cylinder heads were professionally rebuilt and completed at the end of November 2003. At the beginning of December 2003, the owner delivered the original Hawaii Motor Rebuilders short block to Luke's Custom Machine & Design with the intention of rebuilding it. However, a careful inspection of the engine block revealed a crack leading from the deck surface to a cylinder wall and a very serious crack between the center main bearing web and the oil pan rail. Unfortunately, that engine block could not be used.

The blown Ardun engine described in this section of the book was built by Luke's Custom Machine & Design, completed in August 2004, and installed in a 1940 Mercury coupe. The owner's father purchased this vehicle in 1941 and sold it to him in 1957 for $1. This one-family vehicle has approximately 275,000 original miles on it. The Ardun is the ultimate engine for this truly classic automobile.

Over the past year, I have had the pleasure of dealing with a fine gentleman by the name of E. Dean Butler. Dean and his brother Don are the owners and managers of Zakira's Garage (www.zakiras.com) in Cincinnati, Ohio. Their shop restores, maintains, and races vintage automobiles. They also reproduce Ford Model T and Mercer T-head engine blocks and (Harry) Miller carburetors.

Dean owns a 1950 Allard J2 automobile that was originally powered by an Ardun V-8 engine equipped with Hilborn fuel injection. Its current Ardun engine is fitted with three Stromberg 48 carburetors and Speed Products air horns producing 325 bhp (brake horsepower). Dean also races a 1953 Allard J2 originally owned by the late General Curtis LeMay. Along with the two Ardun V-8 engines, Dean has two Ford V-8-60 engines equipped with the extremely rare Ardun OHV conversion kit.

Dean Butler built his first Ardun V-8 engine in 1974 with assistance from Tom Senter, Clem Tebow, Cotton Werksman, and Zora Arkus-Duntov, whom he knew personally. Dean met Luke Balogh at Tom Senter's house in California in the early 1980s. Tom Senter's blown Ardun V-8 engine, featured on the cover of the June 1971 issue of *Rod & Custom* magazine, shows a Vertex magneto attached to a Vertex angle drive adapter. That adapter was given to Dean by the famous drag racer Charlie Bang and then loaned to Tom Senter.

Over the years, Dean Butler had many discussions with Zora Duntov about the history of the Ardun V-8 and V-8-60 conversion kits. Duntov explained that the Talbot engine was the inspiration for the Ardun conversion and that he knew Walter Becchia, who patented the Talbot design in 1934. Duntov showed Dean a copy of the patent. The original intention of the Ardun engine was for use in trucks, although Duntov was never able to sell the concept. The headline on the front page of the business section in the Monday, August 18, 1947, edition of the *New York Times* confirms this, stating "New Truck Engine Nears Completion." Duntov informed Dean that he only learned of the similarity between the Ardun V-8 cylinder heads and the early Chrysler Hemi cylinder heads when he returned to the United States after working at the Allard Motor Company in the United Kingdom. None of the major automobile manufacturers had shown any previous interest in Duntov's cylinder head design.

Duntov mentioned to Dean that only 20 conversions kits were manufactured for the Ford V-8-60 engine. A promoter ordered them as a cheaper alternative to the small Offenhauser engines used in midget race cars, and they were intended for a series at the Polo Grounds. That series was not developed and, as a result, the promoter never took delivery of the V-8-60 conversion kits. Duntov subcontracted the design and foundry work for the V-8-60 cylinder heads to a company named Veritas, in what became East Germany. At the time, Veritas was manufacturing engines built to a pre–World War II design of the BMW 328 engine. That hemispherical cylinder head engine had transverse pushrods between the rocker arm shafts on the exhaust side, which is the design used for the V-8-60 engine. One of the V-8-60 Ardun conversion kits that Dean purchased in the United Kingdom came wrapped in East German newspapers. End of history lesson!

Part A

The specifications advertised for the Ardun 1000 V-8-175 OHV conversion kit for a 239.4-ci Ford flathead V-8 engine were 160 horsepower at 4,000 rpm and 225 ft-lb torque at 2,500 rpm. This information was based on a non-modified engine block, 7.0:1 compression ratio, standard Ford flathead V-8 ignition, fuel pump, starting motor, generator, and two-barrel carburetor. The gasoline used was rated at an amazing 67 octane. The maximum power was listed as 175 horsepower at 5,200 rpm.

Purchase and Magnaflux

The best engine block to use as the basis for an Ardun engine is a 1949–1953 Ford flathead V-8 model. I prefer this later generation of engine block to the pre-1949 engine blocks for three good reasons:

i) better cooling design, due to the enlarged size of the coolant flow holes at the rear of the engine block;
ii) improved rear crankshaft oil seal design;
iii)and there is no integral bell housing. The 1949–1953 Ford flathead V-8 engine blocks are easier to adapt to a wider variety of late-model transmissions.

The common casting number of the 1949–1950 Ford and Mercury flathead V-8 engine blocks was 8BA. The Canadian casting number for those years was C8BA. The common casting number for the 1951–1953 Ford and Mercury flathead V-8 engine blocks was 1BA. The Canadian casting number for those years was C1BA. The only difference I have seen in these Ford and Mercury flathead V-8 engine blocks is that the 1951–1953 Canadian Mercury engine blocks have the 1CM-6658-B cast-steel oil pump idler gear cover. Ardun Enterprises is currently designing an Ardun aluminum engine block.

There were other casting numbers for the Ford flathead V-8 engine blocks. I have heard of a 1949 Ford flathead V-8 engine block with the casting number D1111. The engine block that was used as the basis for the blown Ardun engine has a casting number of C2384. This engine block had been wrapped up and tucked away in the bowels of Luke's Custom Machine & Design for years. (I am convinced the hidden entrance to King Solomon's mine is located somewhere within that shop!) After consulting with a few of the foremost authorities on stock Ford flathead V-8 engines, I have determined this engine block to be a 1949 Ford flathead V-8 with some unique features that are not usually encountered in common 1949–1953 Ford flathead V-8 engine blocks. There are two 1/4-inch NPT water jacket drain plugs (one on each side) at the bottom front of the engine block behind the water pumps. There is a 1/8-inch NPT oil gallery line plug located at the rear of the engine block above the gasket area for the oil pump idler gear cover. The fuel pump pushrod bushing diameter is 0.375-inch i.d., similar to the 1946–1948 Model 59 engine blocks. The water pump bypass hole on each side at the front of the deck surface will accommodate a 1/4-inch NPT plug, not the usual 1/8-inch NPT plug. There is an additional opening in the right (passenger's) side at the back of the valve lifter gallery that permits oil to return to the oil pan. There are only two oil line fitting holes located on the left rear (driver's) side, unlike most 1949–1953 flathead engine blocks that have three oil line fitting holes.

This 1949 Ford flathead V-8 engine block (casting number C2384) was used as the basis for the blown Ardun engine.

Cleaning

Refer to Section I

Porting and Polishing

The intake and exhaust ports in the Ford flathead V-8 engine block were not ported and polished, because they were not used. We installed six Ardun 1203 exhaust manifold block-off plates (where the exhaust manifolds bolt onto the engine block).

Detailing and Valve Lifter Bosses

It is not necessary to drill holes in the valve lifter bosses to adjust the valve lifters. The valves in an Ardun engine adjust at the rocker arms. Also refer to Chapter 1.

Full-Flow Oiling

In order to convert this particular 1949 Ford flathead V-8 engine block to a full-flow oil system, a third hole for the drain-back line must be drilled. The center of the third hole should be 1 inch from the center of the vertical pressure-out hole toward the centerline of the engine block, and this hole is tapped for a 3/8-inch NPT fitting.

The horizontal hole for the oil line fitting on the lower left side near the back of the engine block was used for the pressure-out line; it should be drilled and tapped using a 3/8-inch NPT tap. This hole leads to a horizontal crossover

The notches in the bottom of the cylinder bores are necessary for clearance when using H-beam connecting rods. The underside of the engine block has been painted with Glyptal G-1228A medium grey gloss enamel.

passage inside the engine block. Drill and tap the crossover passage using a 1/4-inch NPT tap, and then install a stainless-steel 1/4-inch NPT plug in the passage using Permatex aviation form-a-gasket. The vertical threaded hole above the horizontal hole is already drilled and tapped for a 1/4-inch NPT fitting. This hole will be used for the pressure-out lines to the rocker arm assemblies and for an oil-pressure gauge.

Relieving the Engine Block

A relieved Ford flathead V-8 engine block cannot be used as the basis for an Ardun engine, as it would be impossible to seal the cylinder heads.

The third oil line fitting hole has been drilled and tapped in the 1949 Ford flathead V-8 engine block. The number 1 flag is for the pressure-out line, the number 2 flag is for the pressure-out lines to the rocker arm assemblies and for an oil-pressure gauge, and the number 3 flag is for the drain-back line.

Heat Risers

The heat riser passages are a nonissue, because they will not be used.

Retap and Chamfer Bolt Holes

Refer to Section I

Redi-Strip the Engine Block

Refer to Section I

Glyptal

Refer to Section I

Machining

Cylinder Boring

The 1949 Ford flathead V-8 engine block was bored and the cylinders honed and deglazed, resulting in a final bore of 3 5/16 inches (3.3125 inches). Purchase the pistons beforehand and deliver them to the machine shop with the engine block. The machine shop will then measure each piston with a micrometer and bore the engine block accordingly to ensure there is 0.006-inch piston-to-bore clearance (after cylinder honing), measured below the bottom of the wrist pin perpendicular to the wrist pin. This is the piston clearance Ross Racing Pistons recommends for its forged aluminum blower pistons. The engine block is usually bored to within 0.003 to 0.004 inch of the final bore and then honed and deglazed, resulting in the final cylinder bore. Also refer to Chapter 1.

Align Hone

Because a blower would be installed on this engine, billet steel number 1 and 2 main bearing caps were fitted, and all the main bearing caps were equipped with new aircraft-quality bolts of at least 180,000 psi. Hawaii Motor Rebuilders fabricated the number 1 and 2 main bearing caps for the original Ardun engine. These excellent main bearing caps are very similar in design to the main bearing caps supplied by Doug King Enterprises. Even though these main bearing caps had been installed on another engine block, they had to be machined to fit the 1949 Ford flathead V-8 engine block. As a result, align boring and honing were also necessary. Also refer to Chapter 1.

The 1949 Ford flathead V-8 engine block has been Redi-Stripped, and the valve lifter gallery and front of the engine block have been painted with Glyptal G-1228A medium grey gloss enamel.

We used this vintage 1946 B.C. Ames Company (Waltham, Massachusetts) Number 30B cylinder gauge to measure the bore and taper of a cylinder.

A Sunnen Horizontal Hone (Model CH-100) align-honing machine or similar machine is used to align hone the engine block. The skill and experience of the operator are crucial to the success of the procedure. Even the best align honing machine, when operated by an unskilled operator, will destroy the engine block. The Sunnen Align Bore Attachment (Model PLB-100) is used with the Sunnen Horizontal Hone (Model CH-100) to align bore an engine block.

Parallel Deck

Refer to Section I

Shot Peen

Refer to Section I

Valve Seat Inserts

The valve seat inserts in the Ford flathead V-8 engine block will not be used, so it is not necessary to be concerned with their condition.

We used a Sunnen horizontal hone (Model CH-100) align-honing machine with the Sunnen align-bore attachment (Model PLB-100) to align bore the Ford flathead V-8 engine block.

Check the pushrod-to-valve-guide-boss clearance to ensure the pushrods do not lie against the valve guide bosses, which would lead to the destruction of the pushrods. Long before the final engine assembly, install the crankshaft, camshaft, timing gears, and assembled cylinder heads. Install the intake and exhaust valve lifters with the pushrods for one cylinder and adjust the valve lash. Rotate the engine through 360 degrees and carefully check for adequate pushrod-to-valve-guide-boss clearance. Carry out this procedure for all the cylinders. In most cases, when a street-performance camshaft with a net valve lift of approximately 0.400 inch is installed with the second-gen-

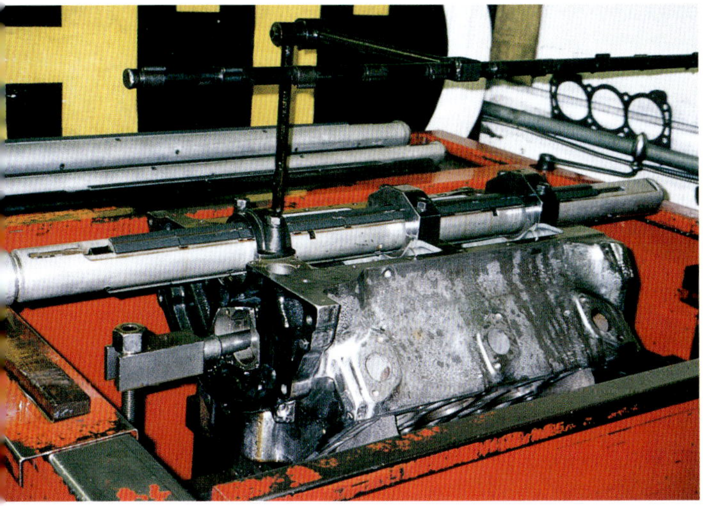

We used the Sunnen horizontal hone (Model CH-100) to align hone the Ardun engine block as well. Notice the number 1 and 2 main bearing caps.

Hawaii Motor Rebuilders professionally fabricated the number 1 and 2 main bearing caps in 1955. All the main bearings caps and bolts have been shot peened.

eration milk-bottle-shape valve lifters, there is sufficient pushrod-to-valve-guide-boss clearance. If there is insufficient clearance, use a high-speed grinder with a carbide bit to remove material from the valve guide bosses.

Valve Grind

The valve grinding of the cylinder heads will be addressed in Chapter 25.

Deglaze

Refer to Section I

Cracks

Due to the fact that an Ardun engine is an expensive proposition, every effort should be made to locate a Ford flathead V-8 engine block without cracks.

Cleaning and Painting

Refer to Section I

Part B

- 1949 Ford 239.4-ci L-head V-8 engine block; casting number: C2384; two-bolt main bearing caps; 24-stud block; stock bore: 3 3/16 inches (3.1875 inches); stock stroke: 3 3/4 inches (3.75 inches); stock compression ratio: 6.8:1; 100 horsepower at 3,800 rpm; 181 ft-lb torque at 2,000 rpm; engine block weight (bare): 200 pounds; **$400**
- Engine block hot tanked and Redi-Stripped; engine block, main bearing caps, and main bearing cap bolts Magnafluxed; **$245**
- All threads retapped; main bearing caps and main bearing cap bolts shot peened; cylinder head bolt holes chamfered; valve lifter bosses deglazed allowing for 0.002-inch valve lifter clearance; new stainless-steel 7/16-inch N.C. socket head set screws (8) installed in top row of cylinder head bolt holes and new stainless-steel 1/4-inch NPT plugs installed in water pump bypass holes at front on each side of deck surface using Permatex aviation form-a-gasket; new stainless-steel water jacket drain plugs (1/4-inch NPT) installed using pipe thread sealant; bottom of cylinder bores notched for connecting rod clearance; underside of engine block, valve lifter gallery, and front of engine block painted with Glyptal G-1228A medium grey gloss enamel; exterior surfaces of engine block sanded and detailed and painted with PPG DGHS 9000 black high-gloss polyurethane; **$777.21**
- Engine block converted to full-flow oil system; **$53.33**
- Engine block bored 1/8-inch (0.125-inch) oversize and cylinders honed and deglazed using Sunnen 800-series 400-grit stones; final bore: 3 5/16 inches (3.31215 inches); piston-to-bore clearance: 0.006 inch measured below bottom of wrist pin perpendicular to wrist pin; **$130**
- Engine block align bored and honed; Hawaii Motor Rebuilders billet steel number 1 and 2 main bearing caps installed with new aircraft-quality cap screw bolts (180,000 psi); **$556.33**
- Engine block parallel decked to 0.010-inch average, below deck; **$120**

Engine Block Total: $2,281.87

CHAPTER 22
CRANKSHAFT

Part A
Purchase

A 1949–1953 Mercury 6303 crankshaft was installed in the Ardun engine. The Mercury flathead V-8 crankshaft that was installed in the original Hawaii Motor Rebuilders Ardun race engine had the main bearing journals grooved. This was a common practice for drag race engines in the 1950s until it was discovered that the procedure did more harm than good. If the oiling holes in the main bearing journals are not cross-drilled (this is the case with flathead crankshafts), grooving the main bearing journals will provide a constant source of oil to the connecting rod bearings. In many OHV engines, only the upper main bearing shell is grooved. The lower half supports a huge load, and this load-bearing surface is reduced dramatically when the bottom main bearing shell is grooved as well. The Clevite 77 and Federal Mogul flathead main bearings are fully grooved on the upper and lower halves, thus negating any meaningful reason for the main bearing journals to be grooved.

Magnaflux

Refer to Section I

Shot Peen and Plug Removal

Refer to Section I

Straighten, Chamfer, and Polish

Refer to Section I

Regrinding and Cleaning

The connecting rod journals were reduced in diameter to 2.00 inches in order to accommodate the SCAT Enterprises 2-239-7000-2000 H-beam connecting rods. In order to take advantage of this situation, the connecting rod journals were offset ground 0.0625 inch and then reduced to the 2.00-inch diameter, resulting in a 4 1/8-inch (4.125-inch) stroke. This stroke, combined with the 3 5/16-inch (3.3125-inch) bore, gives a displacement of 284.5 ci (4.7 liters). The estimated cost to offset grind a Ford or Mercury flathead V-8 crankshaft and reduce the

This completed Mercury crankshaft was installed in the Ardun engine. This crankshaft has been Magnafluxed, aligned, shot peened, offset ground, the oil holes chamfered, the journals polished, and the crankshaft balanced.

This set of Clevite 77 MS-109P-20 main bearings (0.020-inch oversize) was installed in the Ardun engine.

connecting rod journals to 2.00-inch diameter is $250. Also refer to Chapter 12.

Main Bearings

The Clevite 77 (Dana Corporation) MS-109P main bearings are currently available in 0.010-inch oversize and 0.020-inch oversize only for the 1949–1953 Ford flathead V-8 engines. The Federal Mogul 646-M main bearings are currently available in standard size, 0.010-inch oversize, 0.020-inch oversize, and 0.030-inch oversize for the 1949–1953 Ford flathead V-8 engines. Also refer to Chapter 2.

Crankshaft Gear and Pulley

Refer to Section II

Part B

- 1949–1953 Mercury 6303 cast alloy steel crankshaft; casting number: K0308; main journal diameter: 2.498 to 2.499 inches; connecting rod journal diameter: 2.138 to 2.139 inches; nonfloating connecting rod bearings; stock stroke: 4.00 inches; crankshaft weight: 67 pounds; **$400**
- Crankshaft shot peened, aligned, Magnafluxed; main journals ground 0.020-inch undersize; connecting rod journals offset ground 0.0625 inch and reduced in diameter to 2.00 inches; oil holes chamfered; journals polished; crankshaft balanced; new stroke: 4 1/8 inches (4.125 inches); factory pressed-in connecting rod journal oil plugs replaced with new stainless-steel 3/8-inch NPT plugs using Loctite; **$325.62**
- New Clevite 77 MS-109P-20 main bearings (0.020-inch oversize) installed allowing for 0.002 inch crankshaft clearance and 0.004 inch end play; main bearing cap bolts installed using molykote and torqued to 105 ft-lb; **$117.11**
- New Melling 2701 cast-iron crankshaft helical timing gear (22 teeth) and Ford 6310 oil slinger installed with Ford 357654-S crankshaft gear Woodruff key; Luke's Custom Machine & Design new crankshaft sleeve and one-piece crankshaft hub and pulley (6 1/2 inches diameter, single V-belt) installed with Ford 74153-S crankshaft pulley Woodruff key and Ford 351590-S crankshaft pulley bolt washer with Ford 20639-S crankshaft pulley bolt; crankshaft pulley bolt installed using Loctite and torqued to 50 ft-lb; oil slinger painted with Glyptal G-1228A medium grey gloss enamel; crankshaft pulley painted with PPG DGHS 9000 black high-gloss polyurethane; **$320.84**

Crankshaft Total: $1,163.57

CHAPTER 23
CONNECTING RODS AND PISTONS

Part A
Purchase

The Hawaii Motor Rebuilders original Ardun drag race engine was equipped with a set of 1946–1948 Ford 6200 (Model 29A) full-floating connecting rods. There were no modifications to those connecting rods, indicating how strong the flathead connecting rods actually are. A set of SCAT Enterprises 2-239-7000-2000 H-beam forged steel connecting rods were installed in the Ardun engine. Refer to Chapter 13.

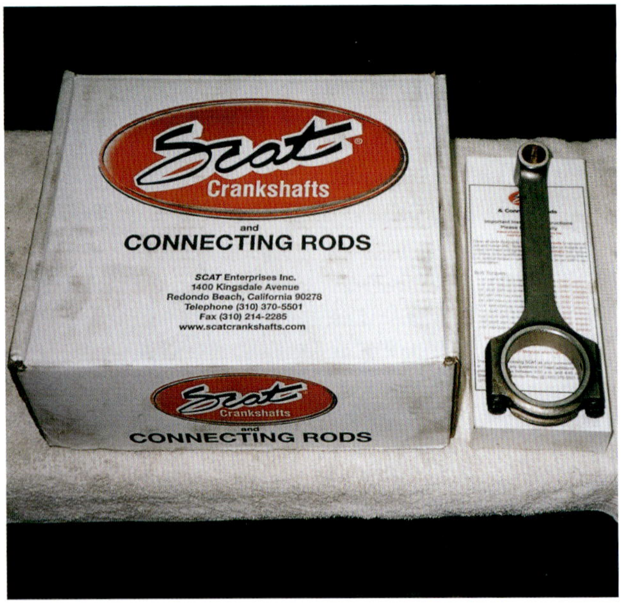

This set of new-design SCAT Enterprises 2-239-7000-2000 H-beam connecting rods had to be reprofiled, so it was not necessary to notch the bottom of the cylinder bores.

Magnaflux and Shot Peen

Refer to Section II

Align, Resize, and Rebush

Refer to Section II

De-Beam

Refer to Section II

Connecting Rod Bearings

Refer to Section II

Pistons

A set of high-dome cast racing pistons was installed in the original Hawaii Motor Rebuilders Ardun drag race engine. The manufacturer of these pistons is unknown; their estimated compression ratio is 13.5:1. Those pistons were not fitted with wrist pin locks; rather, a set of aluminum buttons had been professionally machined to keep the wrist pins in place. This was a nice touch, considering it was done in 1955. A custom set of Ross Racing Pistons forged 2618 T-61 aluminum blower pistons (with a 3 5/16-inch (3.3125-inch) bore, 4 1/8-inch (4.125-inch) stroke, and an 8.0:1 compression ratio) were installed in the Ardun engine. Also refer to Chapter 13.

A cast-aluminum piston (manufacturer unknown) with aluminum wrist pin buttons and a Model 29A full-floating connecting rod were installed in the original Hawaii Motor Rebuilders Ardun drag race engine.

Here is the set of Ross Racing Pistons forged-aluminum blower pistons with an R3315 moly-top ring piston ring set for the blown Ardun engine.

Ross Racing Pistons supplies its own R3315 metric racing piston rings. The top compression rings are moly-coated ductile iron, the second compression rings are cast iron, and the oil rings have a stainless-steel expander. The top compression ring groove width is 1.5 millimeters, the second compression ring groove width is 1.5 millimeters, and the oil ring groove width is 3.0 millimeters. These piston rings are considered file-fit piston rings and they usually require additional gapping. Always adhere to the piston ring manufacturer's specifications for gapping the piston rings. The Ardun engine described here would have a top compression ring gap of 0.017 inch, the second compression ring gap would be 0.021 inch, and the oil ring gap would be 0.017 inch. The Ross Racing Pistons 709-01-21C aircraft-quality chrome-moly steel standard wrist pins with double Spiro-Lox retainers are supplied with the pistons.

After the engine block is parallel decked, and prior to balancing the rotating assembly, check the height of the lower edge of the piston (just below the top) to ensure it is below the height of the deck and does not protrude into the combustion chamber of the cylinder head. This is an extremely critical measurement, and it must be carried out accurately. The lower edge of the top of the piston must be a minimum of 0.010 inch below the deck. If the measurement is less than this, the outer edge (approximately 1/8 inch wide) of the pistons will have to be machined to provide adequate clearance.

Ardun pistons usually have two notches on the dome for valve head clearance, although blower pistons may not. The larger notch is for the intake valve clearance, and it should face upward. The smaller notch is for the exhaust valve clearance, and it should face downward. Remember this when assembling the connecting rods and pistons.

We check the height of the edge of the Ardun piston to ensure it is below the deck surface. The edge of this piston is 0.015 inch below the deck.

Legendary Ardun engine builder Cotton Werksman had this pair of Ardun nitroseal cylinder head gaskets manufactured by Victor/Reinz (GX-13173-1).

Gaskets

The 1949–1953 Ford flathead V-8 engines require a Fel-Pro FS7525B complete gasket set. Many of the gaskets contained in that set will not be used, and it is far more economical to purchase the required gaskets individually from a company such as Jitney Auto Parts. The oil pan gaskets and rear oil seals, timing cover gasket, water pump gaskets, oil pump idler gear cover gasket, and intake manifold gasket will be required.

The Ardun engine requires special cylinder head gaskets, and the best ones to use are those that the legendary Ardun engine builder Cotton Werksman has manufactured by Victor/Reinz (GX-13173-1). These superior nitroseal head gaskets are graphite coated, which allows the aluminum cylinder head to slide on the gasket without experiencing leaks. The only drawback to these gaskets is that they can only be used with a cylinder bore size up to 3 5/16 inches (3.3125 inches). They are available from Luke's Custom Machine & Design or Ardun Enterprises at the current retail price of $100 each.

Part B

- New SCAT Enterprises 2-239-7000-2000 forged 4340 steel H-beam connecting rods, Magnafluxed, shot peened, stress relieved, balanced, and connecting rod caps fitted with dowel alignment sleeves; new ARP 8740 steel (3/8-inch diameter, 190,000 psi) 12-point cap screw bolts installed with molykote and torqued to 42 ft-lb allowing for 0.0061-inch stretch; connecting rod weight: 660 grams each; connecting rod length: 7.00 inches (center to center); connecting rod ratio: 1.70 (with 4 1/8-inch stroke); **$595**
- New Clevite 77 CB-610P connecting rod bearings (standard size) installed allowing for 0.002-inch connecting rod clearance and 0.014- to 0.018-inch side clearance per pair of connecting rods; **$48.61**
- New Ross Racing Pistons forged 2618 T-61 aluminum flat top blower pistons; 8.0:1 compression ratio; 3 5/16-inch (3.3125-inch) bore; piston weight: 339 grams each; new Ross Racing Pistons 709-01-21C aircraft-quality chrome-moly steel straight-wall standard wrist pins installed with new Ross Racing Pistons 4004-J double Spiro-Lox retainers; wrist pin diameter: 0.750 inch; wrist pin length: 2.770 inches; wrist pin weight: 120 grams; piston grooves: top compression ring width: 1.5 millimeters; second compression ring width: 1.5 millimeters; and oil ring width: 3.0 millimeters; new Ross Racing Pistons R3330 ductile moly-top piston ring set installed within manufacturer's recommended arc; top compression ring gap: 0.017 inch; second compression ring gap: 0.021 inch; and oil ring gap: 0.017 inch; new displacement: 284.5 ci (4.7 liters); **$830**
- Complete V-8 engine balance; **$190**
- New Fel-Pro FS7525B complete gasket set installed; **$54.95**

Connecting Rods and Pistons Total: $1,718.56

CHAPTER 24
LUBRICATION SYSTEM

Part A
Purchase

The Ardun engine requires more oil than a Ford flathead V-8 engine to lubricate the rocker arm assemblies. Although it is not mandatory, we installed a Melling M15 high-volume oil pump in the blown Ardun engine. This oil pump does not provide an increase in oil pressure, but it does increase the oil volume. The Melling M15 high-volume oil pump is available from Speedway Motors 577-OP15 at the current retail price of $120. The oil pump is installed with a stainless-steel 5/16-inch N.C. by 1-inch-long hex head bolt and lock washer.

We installed a Melling M15 high-volume oil pump in the blown Ardun engine.

Oil Pan and Pickup

There are three standard types of oil pans available for the 1949–1953 Ford flathead V-8 engines. The most common is the 1949–1953 Ford 6675 center sump passenger car model with a capacity of 3 quarts of oil (4 quarts with the large oil filter). This oil pan would have to be modified to clear a Melling M15 high-volume oil pump. The Ardun engine should have as much oil capacity as possible, which means the 1949–1953 Ford truck 6675 rear sump oil pan is a better choice. This oil pan has a capacity of 4 quarts of oil (5 quarts with the large oil filter), although not every model has a removable bottom inspection plate. Some 1949–1953 Ford truck 6675 oil pans require modification in order to clear a Melling M15 high-volume oil pump.

The best oil pan to use is the 1949–1951 Mercury 6675 rear sump model with an internal baffle. This oil pan has a capacity of 4 quarts of oil (5 quarts with the large oil filter), and it does not have to be modified to clear a Melling M15 high-volume oil pump. All three of the oil pans just mentioned would have to be modified for clearance with a billet steel number 1 main bearing cap, a simple task for any competent machine shop. The oil pan should be Redi-Stripped and mounted using 16 pieces of stainless-steel 5/16-inch N.C. by 3/4-inch-long hex head bolts, lock washers, and AN flat washers. The use of the Ardun cylinder heads will require an additional quart of oil.

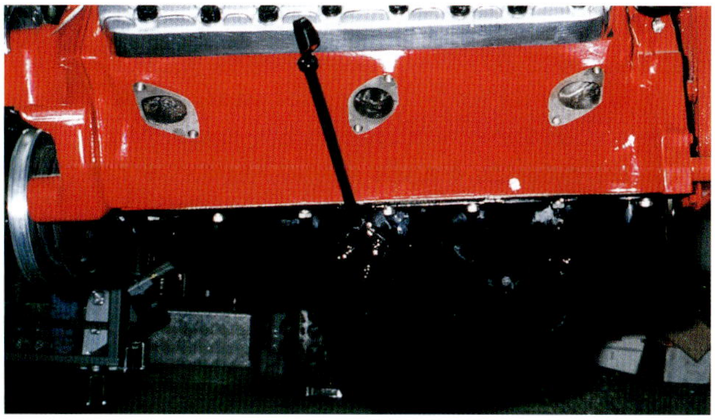

The 1949–1953 Ford 6675 center sump passenger car oil pan has been painted with PPG DGHS 9000 black high-gloss polyurethane.

This photograph shows the internal oil baffle in the 1949–1951 Mercury 6675 oil pan, good for any 1949–1953 Ford flathead V-8 engine.

The 1949–1953 Ford 6754 dip stick and 6750 dip stick tube will protrude outward from the oil pan and interfere with the exhaust tube for the number 7 cylinder on the Ardun engine. The Ford dip stick and dip stick tube are not flexible, so they cannot be bent for clearance. Retap the Ford 6751 dip stick tube boss with a 1/4-inch NPT tap, and install a Weatherhead 202-6 brass adapter (3/8-inch tube to 1/4-inch NPT). Double flare a piece of Summit Racing Equipment 220238 stainless-steel tubing (3/8-inch o.d. by 11 1/4 inches long) on one end and install it with a Weatherhead 100-6 brass inverted flare nut (3/8-inch tube). Reduce a Trans-Dapt 4958 chrome big-block Chevrolet dip stick in length to 15 1/2 inches and install it in the tubing. Notch the "full" level on the dip stick with a hacksaw 2 1/8 inches from the bottom of the dip stick.

The most desirable oil pump cover and inlet tube (oil pump pickup) for the rear sump oil pans is the 1949–1951 Mercury 6615 model with an attached baffle. The second best is the 1949–1953 Ford truck 6615 model. Use the longer 1949–1953 Ford 6615 oil pump cover and inlet tube with the 1949–1953 Ford 6675 center sump oil pans. Jitney Auto Parts is a good source for all flathead oil pans and oil pump pickups.

Install drain line fittings for the cylinder heads and exhaust port block-off plates in the oil pan. One 3/8-inch NPT fitting is welded near the center of the oil pan on each side near the oil pan rail for the cylinder head drain lines. One 1/8-inch NPT fitting is welded near the center of the oil pan on each side near the oil pan rail for the exhaust port block-off plate drain lines. Make certain these oil line fittings are positioned so as not to obstruct the installation of any oil pan bolt.

Ardun Enterprises recently introduced an absolutely beautiful 356 T-6 finned aluminum oil pan for 1949–1953 Ford flathead V-8 engine blocks. The sump is almost full length with an 8-quart oil capacity, an aluminum windage tray, and a cylinder head drain line fitting on each side of the oil pan. The current retail price is $1,295. Provided there is sufficient frame and steering clearance, this polished oil pan would look spectacular on any Ardun engine.

Oil Pump Idler Gear

Some 1949–1953 Ford flathead V-8 engine blocks have the oil pump idler gear installed when they are purchased. Remove the oil pump idler gear prior to hot tanking and Redi-Stripping the engine block. Unlike the earlier models, the 1949–1953 Ford flathead V-8 oil pump idler gear is separate from the oil pump idler gear cover. After removing the oil pump idler gear cover, you will see a 3/8-inch N.C. threaded hole in the face of the oil pump idler gear. Insert a bolt approximately 1 1/2 inches long with a thick flat washer in the hole, attach the jaws of a slide hammer to the flat washer, and pull the oil pump idler gear out of the engine block. You will seldom encounter a damaged oil pump idler gear, and the oil pump idler gear assembly is a readily available, inexpensive item.

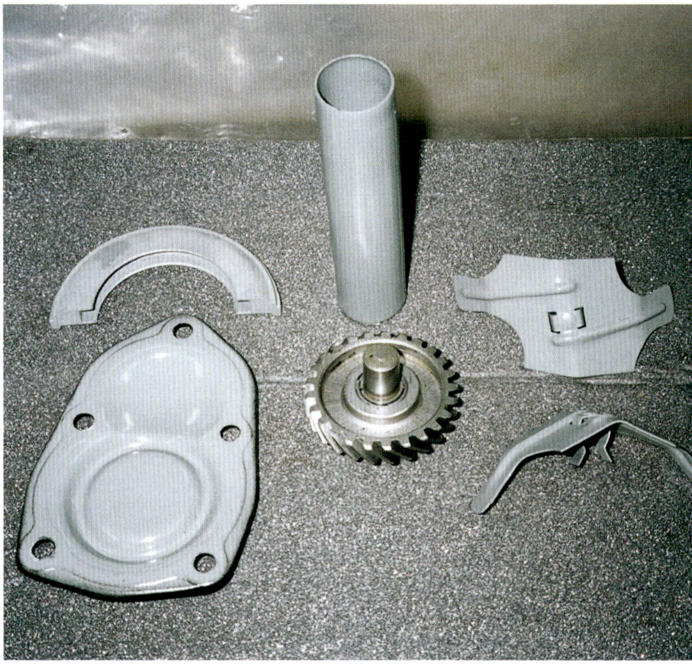

A 1949–1953 Ford 6655 oil pump idler gear and 6656 shaft and a Ford 8BA-6335 rear upper oil seal retainer are shown here. The Ford 6756 crankcase breather tube, Ford 6524 baffles, and Ford 6658 idler gear cover have been painted with Glyptal G-1228A medium grey gloss enamel.

Fuel Pump Pushrod Bushing

Remove the 1949–1953 Ford 6025 fuel pump pushrod bushing prior to hot tanking and Redi-Stripping the engine block. First remove the rear camshaft bearing, and then knock the fuel pump pushrod bushing downward using a brass drift and a hammer. There is a hole in the fuel pump pushrod bushing that must be brazed closed when using an electric fuel pump. Otherwise, there would be a loss of oil pressure if the fuel pump pushrod were not used. If you will use the stock fuel pump, line up the hole in the fuel pump pushrod bushing with the hole in the engine block in order to lubricate the fuel pump pushrod when it is reinstalled.

In order to block off the fuel pump pushrod bushing in this particular 1949 Ford flathead V-8 engine block, have a 0.380-inch-diameter aluminum plug machined and installed using Permatex aviation form-a-gasket.

This is a 1949–1953 Ford 6025 fuel pump pushrod bushing. Although a little difficult to see, the hole in the pushrod bushing has been brazed closed.

Baffles and Breather Tube

The two valve lifter gallery oil-return hole baffles and the valve lifter gallery crankcase breather tube used in 1949–1953 Ford flathead V-8 engines are the same as those used in pre-1949 Ford flathead V-8 engines. These items are easy to locate, and they should cost little or nothing. Refer also to Chapter 4.

Oil Filter

The 1949–1953 Ford flathead V-8 engine blocks have a 1/8-inch NPT fitting hole in the rear left (driver's) side that was used for the drain line from the factory oil filter. Because the blown Ardun engine has been converted to a full-flow oil system, this hole must be plugged with a stainless-steel 1/8-inch NPT plug. Also refer to Chapter 4.

Rocker Arm and Pushrod Lubrication

Have an oil line lubrication system fabricated for the rocker arm assemblies. Luke's Custom Machine & Design fabricates four collars, one for each of the two cylinder head/valve cover studs on each side of the engine. The collars slide over the cylinder head/valve cover studs and are secured in position with socket head set screws. Brass 1/8-inch-diameter tubing passes through the collars and is supported at one end by a stainless-steel bracket. In order to lubricate the ends of the pushrods and rocker arms, brass spray bars with 0.080-inch-diameter oil holes are brazed to the main oil line at the appropriate point for each pushrod and rocker arm. The end of the tubing at the stainless-steel support bracket is brazed closed, and the other end is attached to the oil feed line. This is a very simple, neat, and functional oiling system.

In order to supply oil to the feed lines in the cylinder heads, 3/16-inch-diameter stainless-steel tubing is plumbed from the pressure-out fittings at the rear of the engine block to a T valve at the rear of each cylinder head. The T valve restricts the oil flow to the rocker arm assemblies. Forty-four brass Weatherhead fittings were used with the oil feed lines. There is a drain hole in the bottom center of each cylinder head that permits the oil in the cylinder heads to return to the oil pan. Stainless-steel 3/8-inch-diameter tubing is used for the drain lines from the cylinder heads to the upper area of the oil pan.

Exhaust Port Drain Lines

Ford 6510 valve guide bushings are not installed in an Ardun engine. The absence of valve guide bushings causes oil to accumulate in the exhaust ports of the Ford flathead V-8 engine block behind the Ardun 1203 exhaust port block-off plates. In order to rectify the problem, have a system of drain lines from the exhaust port block-off plates to the upper area of the oil pan fabricated. Drill a hole in each exhaust port block-off plate opposite the lowest point in each exhaust port in the engine block. Tap the hole and install a 90-degree brass 1/8-inch NPT elbow in each exhaust port block-off plate using pipe thread sealant. Plumb stainless-steel tubing from each exhaust port block-off plate fitting to a center fitting, and from there to the upper area of the oil pan. The installation of the exhaust port drain lines required 37 brass fittings. The Summit Racing Equipment double-annealed stainless-steel tubing only comes in 20-foot coils; however, approximately 6 feet of 3/16-inch-diameter tubing was required for the oil feed lines to the cylinder heads and the exhaust port block-off plate drain lines.

Part B

- New stainless-steel 3/8-inch NPT front oil gallery line plug, new stainless-steel 1/8-inch NPT rear oil gallery line plug, and new stainless-steel 1/8-inch NPT engine block oil filter line plug installed using Permatex aviation form-a-gasket; fuel pump pushrod bushing blocked off with new aluminum plug (0.380-inch diameter) using Permatex aviation form-a-gasket; new stainless-steel 1/4-inch NPT water jacket drain plugs installed using pipe thread sealant; 1949–1953 Ford 6657 oil pump idler gear bushing installed with Ford 6655 oil pump idler gear and Ford 6656 oil pump idler gear shaft allowing for 0.002 inch clearance; 1949–1953 Ford 6658 oil pump idler gear cover installed with new stainless-steel hex head bolts and lock washers using Loctite and torqued to 15 ft-lb; oil pump idler gear cover gasket installed using silicone sealant; inside of oil pump idler gear cover, Ford 6756 valve lifter gallery breather tube, and Ford 6524 valve lifter gallery oil-return hole baffles (2) painted with Glyptal G-1228A medium grey gloss enamel; **$122.30**
- New Melling M15 high-volume oil pump (for 1949–1953 Ford) installed with 1949–1951 Mercury 6615 oil pump cover and inlet tube, Ford 6623 oil pump cover

screen, and Ford 6628 oil pump cover screen spring retainer; oil pump end clearance: 0.00025 inch; oil pump cover and inlet tube gasket installed using silicone sealant with bolts torqued to 80 in-lb using Loctite; 1949–1953 Ford 8BA-6335 rear upper oil seal retainer installed using silicone sealant; rear oil seal packings (Ford 6700) soaked in SAE 30-weight motor oil and installed; new Speedway Motors 912-S12853 front one-piece oil seal installed using silicone sealant; **$250.35**

- 1949–1951 Mercury 6675 oil pan (5-quart capacity) installed with new stainless-steel hex head bolts, lock washers, and AN flat washers using Loctite and torqued to 15 ft-lb; oil pan gaskets installed using silicone sealant; oil pan hot tanked, Redi-Stripped, and painted with PPG DGHS 9000 black high-gloss polyurethane; new 3/8-inch NPT fittings installed for cylinder head drain lines, and new 1/8-inch NPT fittings installed for exhaust port block-off plate drain lines; Luke's Custom Machine & Design new custom dip stick tube fabricated using Summit Racing Equipment 220238 polished stainless-steel tubing (3/8-inch o.d. by 0.028-inch wall by 11 1/4 inches long); new Weatherhead 202-6 polished brass adapter (3/8-inch tube to 1/4-inch NPT); new Weatherhead 100-6 polished brass inverted flare nut (3/8-inch tube); fittings installed using pipe thread sealant; new Trans-Dapt 4958 chrome dip stick (15 1/2 inches long) installed; engine lubricated with 6 quarts Pennzoil HD-30-weight motor oil; **$270.59**
- New Trans-Dapt 1045 polished remote oil filter bracket with horizontal inlet/outlet; new Morosso 22400 chrome oil filter; new Trans-Dapt 1032 neoprene rubber oil filter hoses (24 inches long by 1/2 inch NPT) installed with new Weatherhead 48-8 polished brass compression fittings (1/2-inch tube to 3/8-inch NPT) using pipe thread sealant; new Weatherhead 3330-6 polished brass 3-inch male extension (3/8-inch NPT); new Weatherhead 3300-6 polished brass female coupling (3/8-inch NPT); new Weatherhead 3400-6 polished brass 90-degree elbow (3/8-inch NPT); and new Weatherhead 3400-2 polished brass reducer (1/8-inch NPT to 1/4-inch NPT) installed using pipe thread sealant; hoses covered with new Spectre 3001A black nylabraid and new Spectre 3308B chrome champ-clamps; **$96.94**
- Cylinder head oil feed lines fabricated using: new Weatherhead 3700-2 polished-brass T fittings (1/8-inch NPT), new Weatherhead 3325-2 polished-brass hex nipple, new Weatherhead 100-3 polished-brass inverted flare nuts (3/16-inch tube), new Weatherhead 3750-2 polished-brass street T fitting (1/8-inch NPT), new Weatherhead A6845S polished-brass compression angle T valves (3/16-inch tube to 1/8-inch NPT), new Weatherhead 402-3 polished-brass male elbow (3/16-inch tube to 1/8-inch NPT), new Weatherhead 202-3 polished-brass male connector (3/16-inch tube to 1/8-inch NPT), new Summit Racing Equipment 220236 polished stainless-steel tubing (3/16-inch o.d. by 4 feet long) pressure-out lines to cylinder heads; all fittings installed using pipe thread sealant; **$44.34**
- Pushrod/rocker arm lubrication system fabricated using: new Weatherhead 41-3 polished-brass compression fittings (3/16-inch tube), new Weatherhead 60-3 brass ferrules (3/16-inch i.d.), new brass tubing (1/8-inch o.d. by 5 feet long) installed with new stainless-steel end support brackets; all fittings installed using pipe thread sealant; **$479.72**
- Cylinder head oil drain lines fabricated using: new Weatherhead 3220-4-2 polished-brass adapters (1/4-inch NPT to 3/8-inch tube), new Weatherhead 100-6 polished-brass inverted flare nuts (3/8-inch tube), new Summit Racing Equipment 220238 polished stainless-steel tubing (3/8-inch o.d. by 3 feet long); all fittings installed using pipe thread sealant; **$7.43**
- Ardun 1203 exhaust port block-off plates painted with PPG DGHS 9000 black high-gloss polyurethane and installed with new polished-stainless steel hex head bolts and lock washers using Loctite and torqued to 15 ft-lb; exhaust port block-off plate gaskets installed using high-temperature silicone sealant; exhaust port block-off plate oil drain lines fabricated using: new Weatherhead 402-3 polished-brass 90-degree male elbows (3/16-inch tube to 1/8-inch NPT), new Weatherhead 3326-2 polished-brass close nipples (1/8-inch NPT), new Weatherhead 3300-2 polished-brass female couplings (1/8-inch NPT), new Weatherhead 3700-2 polished-brass T fittings (1/8-inches NPT), new Weatherhead 3750-2 polished-brass street T fittings (1/8-inch NPT), new Weatherhead 202-3 polished-brass male connectors (3/16-inch tube to 1/8-inch NPT), new Weatherhead 100-3 polished-brass inverted flare nuts (3/16-inch tube), new Weatherhead 3400-2 polished-brass 90-degree street elbows (1/8-inch NPT), new Summit Racing Equipment 220236 polished stainless-steel tubing (3/16-inch o.d. by 4 feet long); all fittings installed using pipe thread sealant; **$90.67**

Lubrication System Total: $1,362.34

CHAPTER 25
CAMSHAFT AND CYLINDER HEADS

Camshaft Bearings

We installed a set of Clevite 77 SH-21S camshaft bearings in the blown Ardun engine. Also refer to Chapter 5.

Camshaft

Once again, Crower Cams & Equipment Company came through with a superb camshaft for the blown Ardun engine. The camshaft supplied has an advertised duration of 277 degrees intake and exhaust; duration at 0.050-inch lift: 222 degrees intake and exhaust; lobe separation angle: 112 degrees; net valve lift (with 1.20 ratio rocker arms): 0.386-inch intake and 0.384-inch exhaust; valve lash (hot): 0.010-inch intake and 0.012-inch exhaust. The price for this camshaft is outright with no core exchange. Also refer to Chapter 15.

Crower Cams & Equipment Company custom ground this camshaft for the blown Ardun engine. End gears are not supplied with the camshaft.

Camshaft Interchange

Refer to Section I

Degree Camshaft

Refer to Section I

Valve Lifters

The Ardun Engine Company produced two types of solid lifters (tappets or cam followers) for the Ardun engine. The first design was similar in size to the stock Ford flathead V-8 solid lifters except the upper 20 percent of the lifter was reduced to approximately half the diameter. Due to the shorter length, the first design solid lifter usually introduced pushrod-to-valve-guide-bushing-boss interference when used with aftermarket camshafts.

Shadbolt Cams, a legendary local camshaft-grinding company, refaced this set of Ardun 1024 milk-bottle-shaped solid lifters.

Ardun 1024 milk-bottle-shaped solid lifters were the second design. These valve lifters were manufactured using high-tensile nickel-chromium molybdenum steel. They are nonadjustable, 3.260 inches long, weigh 100 grams each, and are the same 1.00-inch diameter as the Ford 6500 valve lifters. These valve lifters are of good quality and will be used in the blown Ardun engine. None of the major U.S. camshaft manufacturing companies offer replacement valve lifters for the Ardun engine; however, they are available from Ardun Enterprises. Most likely, the Ardun valve lifters have been used, and for a cost of approximately $70,

Each Ardun Enterprises milk-bottle-shaped solid lifter is 1.00 inch in diameter, 3.260 inches long, and weighs 121 grams. The Ardun 1024 milk-bottle-shaped solid lifters have the same dimensions but weigh 21 grams less.

The Ardun 1062 intake valves and 1063 exhaust valves were polished in a lathe and are ready for a Serdi valve grind.

any reputable camshaft company can reface them. Carefully inspect the Ardun 1024 milk-bottle-shaped valve lifters for cracks in the pushrod seat cup area.

Valves and Valve Seat Inserts

The Ardun 1062 intake valves were referred to by their "tulip" shape, and they had a 1 7/8-inch (1.875-inch) head diameter, 3/8-inch (0.375-inch) stem diameter, 4.875-inch overall length, and they weighed 122 grams each. The 1 7/8-inch (1.875-inch) head diameter is actually larger than it should be for Ardun 1028 intake valve seat inserts. It was a common practice in the 1950s and 1960s to reduce the Ardun 1062 intake valve head diameter to 1 3/4 inches (1.75 inches). The Ardun 1062 intake valves that we installed in the blown Ardun engine have a 1 3/4-inch (1.75-inch) head diameter, and they have been reduced to 4.860 inches long overall with a resulting weight of 114 grams each.

The Ardun 1063 exhaust valves were also a tulip shape, and they were referred to as "austenitic," which means "a nonmagnetic solid solution of ferric carbide or carbon in iron, used in corrosion-resistant steel." This was named after Sir William Chandler Roberts-Austen (1843–1902). The exhaust valves had a 1 1/2-inch (1.50-inch) head diameter, 3/8-inch (0.375-inch) stem diameter, 4.895-inch overall length, and they weighed 103 grams each. The Ardun 1063 exhaust valves that we installed in the blown Ardun engine have a 1 1/2-inch (1.50-inch) head diameter, and they were reduced to 4.865 inches long overall with a weight of 101 grams each.

The Ardun 1062 intake valves and Ardun 1063 exhaust valves, originally installed in the Ardun cylinder heads built by Hawaii Motor Rebuilders, were in excellent condition. All the valves were completely polished in a lathe and then sent off to High Performance Engines for a Serdi-machined multiangle blueprint valve grind. Ardun Enterprises sells replacement stainless-steel valves for the Ardun cylinder heads.

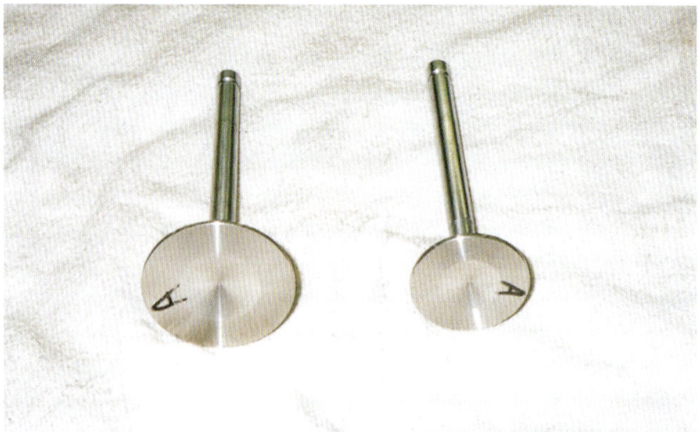

The Ardun Enterprises stainless-steel intake valve is 1 7/8 inches in diameter, 11/32 inch stem, 4.915 inches long, and weighs 123 grams. The stainless-steel exhaust valve is 1.50 inches in diameter, 11/32 inch stem, 4.935 inches long, and weighs 98 grams.

The Ardun aluminum cylinder heads were fitted with Ardun 1028 phosphor bronze intake valve seat inserts and Ardun 1029 phosphor bronze exhaust valve seat inserts. In the past, there were a few reported instances of valve seat inserts coming loose as a result of the different expansion rates between the bronze valve seat inserts and the aluminum cylinder heads. A professional machine shop usually inspects the valve seat inserts prior to performing a valve grind and ensures the valve seat inserts are securely in place. The valve seat inserts in the Hawaii Motor Rebuilders Ardun cylinder heads were in very good condition.

The Serdi-machined multiangle blueprint valve grind has been performed on the cylinder heads for the blown Ardun engine. The valve seat inserts and valve seats are beautifully blended into the ports.

As previously mentioned in this book, a Serdi-machined multiangle blueprint valve grind is the ultimate in valve grinds. It is extremely important to blend the valve seat inserts in with the valve seats and finally with the ports themselves. This procedure is also referred to as "pocket porting" and is a sure way to pick up additional horsepower. A Serdi valve grind guarantees a perfect blend in the ports. The estimated cost for a Serdi-machined multiangle blueprint valve grind for a pair of OHV cylinder heads is $135.

The Ardun 1062 intake valves and Ardun 1063 exhaust valves do not have hardened tips. It is a good practice to install lash caps on these types of valves, as they will protect the ends of the valves from excessive wear and increase the rocker-arm-to-valve-spring-retainer clearance. We installed a set of Manley 42108-16 lash caps (3/8-inch stem diameter valves and a 0.220-inch minimum tip thickness) in the blown Ardun engine.

Valve Guides

The Ardun aluminum cylinder heads were fitted with Ardun 1013 phosphor bronze intake valve guides and Ardun 1014 phosphor bronze exhaust valve guides. The valve guides are pressed into the cylinder head and should be knurled to ensure they stay securely in position in the cylinder head. The valve guides in the original Hawaii Motor Rebuilders Ardun cylinder heads had excessive valve-to-valve-guide clearance. As a result, we installed new K-Line KL1842STA universal length 3/8-inch (0.375-inch) diameter bronze bullet valve guide liners. The proper valve-to-valve-guide clearance in a street-performance engine using today's unleaded gasoline is 0.0015 to 0.002 inch.

The Ardun 1013 bronze intake valve guide was knurled to ensure it remains securely in position in the cylinder head.

Valve Springs

Due to the fact that we installed a blower on the Ardun engine described here, valve springs must be installed with sufficient tension to prevent the intake valves from being pulled off the valve seat inserts when the engine experiences boost. Valve spring selection for Ardun cylinder heads is limited by the outside diameter of the valve spring, the installed height, the coil bind, and the possibility of contact between the rocker arm and valve spring retainer. If coil bind is experienced with valve springs, the situation will cause fatigue and eventually break the valve springs. A minimum of 0.060 inch (preferably 0.100 inch) clearance between the valve spring coils in the fully open position is required. A set of Crane Cams 99480-16 aircraft-quality, silicon chrome, single helical coil valve springs with damper will be used. The installed height is:

Installed height, valves closed: 90 pounds at 1.578 inches

Installed height, valves open: 260 pounds at 1.200 inches (0.400-inch lift)

1.460-inch o.d. and 1.070-inch i.d.

Coil bind: 0.935 inch

Color: red/olive green

Maximum lift: 0.510 inch

Install the valve springs in the Ardun cylinder heads using Ardun 1071 valve spring seat cups, or valve spring washers, as the Ardun Engine Company referred to them. The use of valve spring seat cups or valve spring seat discs is mandatory with aluminum cylinder heads in order to prevent the cylinder heads from galling.

This set of Ardun 1071 valve spring seat cups has been glass beaded and sprayed with graphite coating.

Iskenderian Racing Cams 6005 dual silicon valve springs with damper and matching 507-ST chrome-moly steel valve spring retainers are used with the Ardun Enterprises cylinder heads.

Valve Spring Retainers, Locks, and Seals

Use the Crane Cams 99914-16 (see Appendix B: Resources) heat-treated chrome-moly steel 7-degree valve spring retainers with Crane Cams 99480-16 valve springs and 3/8-inch stem diameter valves. If those valve spring retainers are not readily available, the Iskenderian Racing Cams 3607-ST heat-treated chrome-moly steel 7-degree valve spring retainers can be used with Crane Cams 99480-16 valve springs and 3/8-inch stem diameter valves. The Iskenderian Racing Cams 3607-ST valve spring retainers are 0.060 inch higher than stock; the outer step diameter is 1.065 inches and the inner step diameter is 0.730 inch.

Upgrade the stamped-steel valve stem locks in any Ardun engine. We installed a set of Crane Cams 99098-1 machined 4140 chrome-moly heat-treated steel 7-degree valve stem locks with a black-oxide finish and standard height in the blown Ardun engine. Crane Cams products are available at most speed shops.

The Ardun cylinder heads were missing valve stem seals—obviously not an area of concern in the early 1950s.

These blown Ardun cylinder heads are polished and fully assembled. Crane Cams 99480-16 valve springs, Iskenderian Racing Cams 3607-ST valve spring retainers with Crane Cams 99098-1 valve stem locks, and Enginetech S2890 valve stem seals (orange color in lower part of valve springs) are visible.

Some engine builders do not install valve stem seals on the exhaust valves in engines equipped with hemi cylinder heads. It is a good practice to install valve stem seals on all the valves, and the best valve stem seals for street-performance engines are the Enginetech S2890 viton positive-seal model for use with 3/8-inch stem diameter valves. The valve guides do not usually have to be machined in order to install these valve stem seals.

Camshaft Gear and Timing Cover

Never use a Ford flathead V-8 fiber camshaft timing gear in an Ardun engine. The added stress of the overhead valvetrain will strip the teeth off the fiber gear. We installed a 1949–1953 Ford flathead V-8 camshaft with the appropriate distributor and timing gear cover, which requires the use of later model timing gears, such as the Melling 2702 aluminum camshaft timing gear. Also refer to Chapter 15.

We installed a Mallory distributor in the blown Ardun engine, requiring a timing gear cover with a boss to support the end of the distributor shaft. As a result, we installed a 1951–1953 Ford C8RT-6059-B/COBA-6059-A cast-aluminum timing gear cover. Also refer to Chapter 17.

One-Piece Oil Seal

Refer to Section I

Cylinder Heads

The Ardun 1001 OHV hemispherical combustion chamber cylinder heads were cast with Alcoa 355 T-6 aluminum alloy. These cylinder heads have a combustion chamber volume of 88 cc and weigh 30 pounds (bare) each. There were two types of cylinder heads produced. The first design has an Ardun 1018 aluminum water offtake (water outlet elbow) located at the front of the cylinder head, and the second design has an Ardun 3402 aluminum water outlet elbow located at the top center of the cylinder head between the two middle intake ports. The Ardun article by Jay Fitzhugh in Volume 25 of the *Rodder's Journal* reports that the serial numbers on the cylinder heads range from 160 to 1040. If there were only 200 sets (or even 400 sets) produced, the serial numbers cannot be sequential. The Ardun Engine Company advertised the complete Ardun 1000 overhead valve (OHV) conversion kit for the Ford flathead V-8 (V-8-100) at $479 on March 1, 1950.

There are probably a number of Ardun cylinder heads being used as wall hangers today because they are cracked and their owners believe they are beyond repair. Luke's Custom Machine & Design has had great results with the repair of damaged Ardun cylinder heads. In the past couple of years, Luke Balogh has resurrected three pairs of seriously

Cotton Werksman sent these second-design Ardun cylinder heads (947 and 1039) to Luke's Custom Machine & Design to repair the damaged area around the welds on the top surface. The welding was a failed attempt by an aircraft company to fix a number of cracks.

This material had to be removed from an Ardun cylinder head in order to properly repair it. The damage extended into the threads of one of the spark plug tube holes, making this a very difficult and time-consuming job.

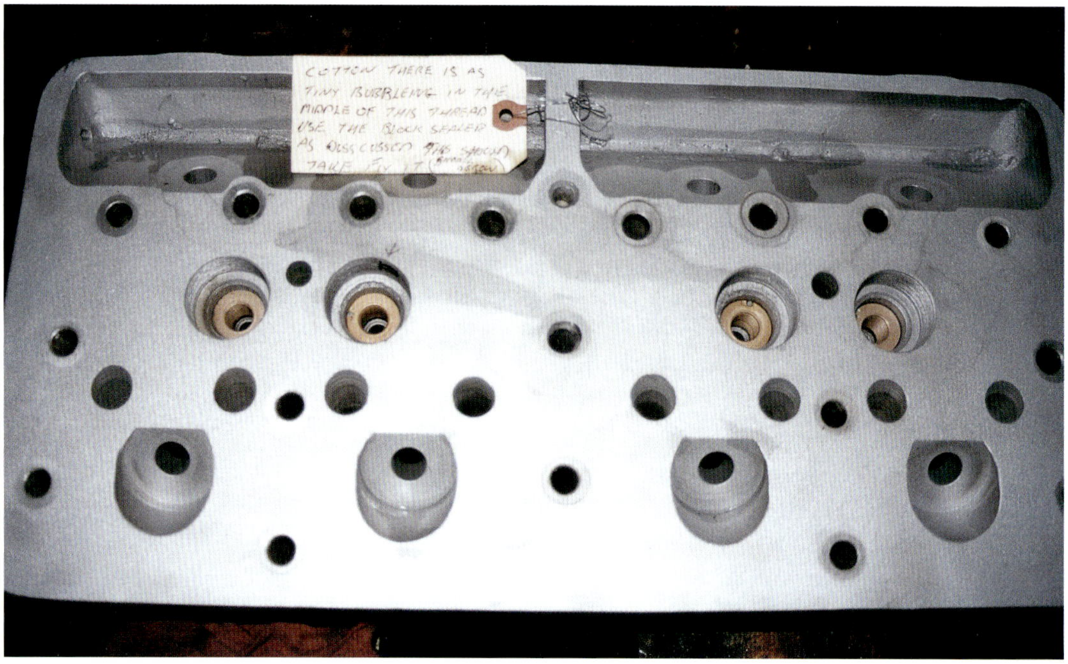

Luke's Custom Machine & Design completed the repairs to one of Cotton Werksman's Ardun aluminum cylinder heads.

damaged Ardun cylinder heads that would otherwise have been consigned to the scrap heap. One of the problems encountered while repairing Ardun cylinder heads is the porosity of the 355 T-6 aluminum alloy used in the manufacture. The residue from inside the cylinder head has a tendency to bubble through the aluminum when it is heated, which hinders the welding. As one crack is repaired, another crack develops. It is a very difficult and time-consuming matter to properly repair Ardun cylinder heads.

Have the Ardun cylinder heads pressure tested to ensure there are no cracks. Luke's Custom Machine and Design has fabricated an aluminum pressure testing plate that bolts to the combustion chamber side of the cylinder head. Compressed air is injected into the cylinder head through the water outlet elbow. The holes in the pressure-testing plate permit the inspection of the combustion chamber for leaks while the cylinder head is under pressure.

Use the intake manifold gaskets as templates to gasket match the Ardun cylinder head intake ports. Grind any casting flash from inside the intake and exhaust ports. Use the exhaust manifold gaskets as templates to gasket match the Ardun cylinder head exhaust ports. You may have the intake and exhaust ports polished, but they should not be enlarged. Use extreme caution when polishing, as the material around the ports is thin in places, particularly around the pushrod bulge. The combustion chamber can be polished, but once again, only remove a minimal amount of material in order not to increase the size of the combustion chambers.

Luke's Custom Machine & Design fabricated this aluminum pressure-testing plate to detect cracks in Ardun cylinder heads.

If there is any grease or oil on the Ardun cylinder heads, clean them with wax and grease remover. Glass bead the cylinder heads to bring them back to the original factory condition. Prior to glass beading the cylinder heads, make certain all the water passage holes are masked off with duct tape, otherwise there will be glass bead material falling out of the cylinder heads and into the engine block for years to come. You may also leave the pressure-testing plate (fabricated by Luke's Custom Machine & Design) in

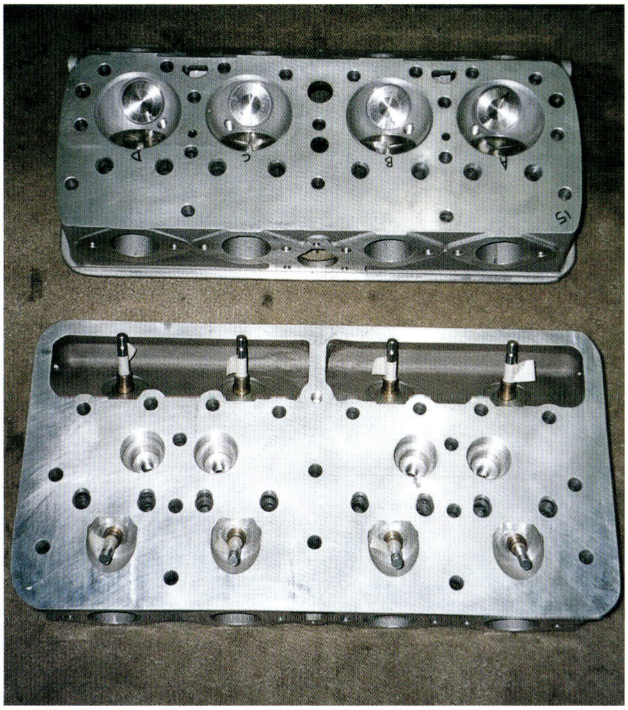

Here is a brand-new pair of Ardun Enterprises 356 T-6 aluminum Ardun cylinder heads. The stainless-steel valves are 1 7/8-inch (1.875-inch) diameter intake and 1 1/2-inch (1.50-inch) diameter exhaust.

Luke's Custom Machine & Design uses a milling machine to surface an Ardun cylinder head.

place for the glass-beading operation. If you really want to get exotic, have someone like "Tim the polishing guy" buff the Ardun cylinder heads—they will come up like chrome. After polishing the cylinder heads, spray them with Eastwood Company 10200Z diamond clear gloss finish or V.H.T. SP-115 clear high-temperature coating to prevent the polished cylinder heads from tarnishing.

Have the Ardun cylinder heads surfaced by a competent machine shop to ensure the cylinder head gasket surface is completely flat. This Ardun engine block was parallel decked and combined with the surfaced cylinder heads; absolutely true cylinder head gasket surfaces are guaranteed.

The Ardun cylinder heads should be cc'd after any rough areas are removed and the final installation of the valves has been carried out. To improve performance, equalize the combustion chamber volume of the cylinder heads by smearing grease on the perimeter of the combustion chamber and placing a plexiglass plate, with a small hole in the top, on top of the grease. Place a glass burette with a stand, capable of holding approximately 150 cc of water above the hole in the plexiglass plate. Drain the water through the hole in the plexiglass plate until the combustion chamber is filled. The amount of water used will be shown on the burette. Perform this exercise for each combustion

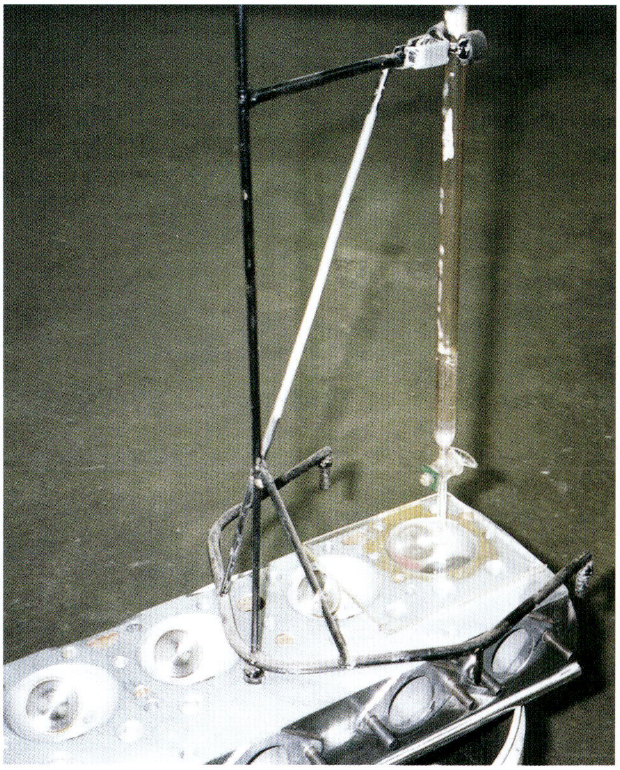

An Ardun cylinder head is being cc'd.

chamber. It might be necessary to grind a small amount of material from the combustion chambers before they are all exactly the same volume. The Ardun cylinder heads described in this section of the book have a final combustion chamber volume of 89 cc. Incidentally, this procedure is applicable to all flathead and OHV cylinder heads. The estimated cost to pressure test and surface a pair of Ardun cylinder heads is $75.

The Ardun cylinder heads can be converted from 18-millimeter to 14-millimeter spark plugs using a Champion J-202 spark plug adapter. Although 18-millimeter spark plugs are still available today, the heat ranges are limited. The 14-millimeter spark plugs, commonly known as "peanut plugs," come in a wider variety of heat ranges. The Ardun engine described here was equipped with 18-millimeter spark plugs.

The tip of the spark plug is recessed in the Ardun cylinder head, creating what is known as a "cartridge fire" situation. Ideally, the spark plug should be lower in the cylinder head so that the spark plug tip is inside the combustion chamber, thereby promoting better flame travel and resulting in a cleaner and smoother-running engine. The spark plugs will benefit from the cooling effect of the incoming (intake) fuel mixture and will actually run cooler at higher engine speeds. The only way to lower the spark plugs in the Ardun cylinder heads is to machine the shoulders where the spark plugs sit to a greater depth. The material in this area of the Ardun cylinder heads is quite thin and there is a good chance of striking water when machining the spark plug shoulders. Unless you have the rare ability to repair Ardun cylinder heads, leave them alone. Removing the spark plug adapter will only lower the spark plug approximately 1/8 inch (0.125 inch) in the cylinder head. The Ardun Enterprises cylinder heads have the spark plugs located so they protrude into the combustion chamber.

Whether the spark plug adapters are going to be used or not, remove them prior to glass beading the Ardun cylinder heads. When the spark plug adapter is installed, there is a pocket below the spark plug adapter and the hole leading to the combustion chamber. This pocket is a great dirt trap and must be cleaned out. Open up the hole leading to the combustion chamber using a 15/32-inch drill bit and chamfering the hole on both sides. Have an experienced machine shop remove the spark plug adapters and open up the holes leading to the combustion chambers. Do not allow anyone that is not an Ardun expert to touch

Luke's Custom Machine & Design used a milling machine to turn this aluminum block into an Ardun-style cylinder head. There are no water jackets in this cylinder head, as it is a racing model.

Ardun cylinder heads. Run an 18-millimeter spark plug tap through all the spark plug holes in the Ardun cylinder heads, and use a 1 1/2-inch N.F. tap to clean up the threads for the spark plug tubes. If the spark plug adapters are going to be reinstalled, clean up the threads with a 14-millimeter tap and install them using antiseize compound.

A few years ago, Luke Balogh decided to manufacture a set of Ardun-style aluminum cylinder heads for his Ford flathead V-8 engine that will be eventually raced at the Bonneville salt flats. These cylinder heads will be strictly used for racing; therefore, water passages were not deemed necessary. The project started with a wooden model of a cross-section of an Ardun cylinder head showing the layout of the valves and combustion chamber. Two solid blocks of 356 T-6 aluminum, a milling machine, years of experience, and approximately 200 hours of labor went into producing a pair of unique hemispherical combustion cylinder heads.

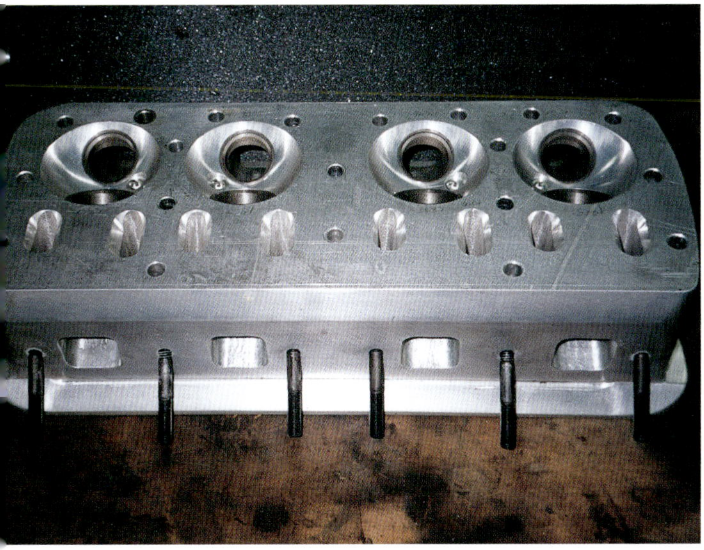

Luke's Custom Machine & Design machined this Ardun-style cylinder head. This picture shows the intake port and the combustion chambers with the spark plugs protruding.

The Ardun Engine Company provided cylinder head studs for use with Ardun cylinder heads. The studs have 7/16-inch N.C. on both ends, unlike most cylinder head studs that have N.C. threads on the engine block end and N.F. threads on the cylinder head end. The Ardun 1005 "short" studs are 4 13/16 inches long. The Ardun 1056 "medium" studs are 5 3/8 inches long. The Ardun 1058 "long" studs are 9 13/32 inches long. The upper 2 3/32 inches of the long studs are used to secure the valve covers. The Ardun 1070 5 13/16-inch studs are also used to secure the valve covers. There are 12 short studs, 24 medium studs, 4 long studs, and 2 valve cover studs. The stock Ardun cylinder head studs are acceptable for the majority of naturally aspirated street-performance engines.

Forty Ardun 1036 hardened flat washers and 40 Ardun 1037 nuts were used on the cylinder head studs. If any of the Ardun 1036 flat washers are damaged or missing, replace them with Manley Performance Parts 42102 heat-treated head bolt washers. None of the cylinder head studs are located outside the valve covers, so chrome acorn nut covers are not required. Six Ardun 1031 flat washers and six Ardun 1015 acorn nuts are used to secure the valve covers. Install the cylinder head studs in the engine block with a liberal amount of Permatex aviation form-a-gasket. Coat the shafts of the cylinder head studs with antiseize compound, and install the cylinder head stud nuts using molykote. Install the Victor/Reinz nitroseal cylinder head gaskets dry, without any cylinder head gasket sealant. The Ardun 1037 cylinder head stud nuts are torqued to 50 ft-lb.

This valve cover side view shows the exhaust ports of the finished Ardun-style aluminum cylinder head fabricated by Luke's Custom Machine & Design.

The Ardun 1055 short studs, Ardun 1056 medium studs, Ardun 1058 long studs, Ardun 1070 valve cover studs, Ardun 1036 flat washers, and Ardun 1037 nuts have all been glass beaded.

If the Ardun engine is going to be equipped with a blower or used for all-out racing, install a top-quality cylinder head stud kit. Ardun Enterprises offers a cylinder head stud kit manufactured by ARP using 8740 chrome-moly steel with a black-oxide finish rated at 190,000 psi. The washers are hardened and parallel ground, and the aerospace-quality nuts are 12-point. There are 12 pieces of 4 3/4-inch-long short studs, 24 pieces of 5 3/8-inch-long medium studs, four pieces of 9 1/2-inch-long long studs, two pieces of 5 3/4-inch-long valve cover studs, 40 washers, 40 nuts, and six brass valve cover acorn nuts included with the kit. Ardun Enterprises includes neoprene rubber O-rings in its cylinder head stud kit when the OHV conversion kit is purchased. The O-rings are installed under the flat washers and will help prevent water leaks. The complete custom-made cylinder head stud kit is available from Ardun Enterprises for the current retail price of $520.

The four cylinder head bolts (on each side) indicate which bolt holes must be plugged on the Ford flathead V-8 engine block when installing Ardun cylinder heads.

After all the machine work on the engine block has been carried out and the engine block has been thoroughly cleaned prior to painting, there are four 7/16-inch N.C. by 1/2-inch-long stainless-steel plugs (on each side of the engine block) that must be installed in the cylinder head bolt holes. The holes that must be plugged in the Ardun engine block are the first top hole from the front of the engine block, the third top hole from the front of the engine block, the fourth top hole from the front of the engine block, and the sixth top hole from the front of the engine block. Install the cylinder head bolt hole plugs using Permatex aviation form-a-gasket.

The water pump bypass hole (one on each side) at the front of the engine block deck surface must be plugged when installing Ardun cylinder heads on a 1949–1953 Ford flathead V-8 engine block. The water pump bypass holes in the majority of 1949–1953 flathead engine blocks can be tapped with a 1/8-inch NPT tap, although some engine blocks require a 1/4-inch NPT tap. Plug the water pump bypass holes with stainless-steel pipe plugs using Permatex aviation form-a-gasket.

Pushrods and Rocker Arms

Ardun 1061 nonadjustable intake pushrods and Ardun 1074 nonadjustable exhaust pushrods are used with the first-design Ardun camshaft lifters. Ardun 1099 nonadjustable intake pushrods and Ardun 1100 nonadjustable exhaust pushrods are used with second-design milk-bottle-shaped camshaft lifters. All the Ardun pushrods were manufactured using high-tensile nickel-chromium molybdenum steel and are a tapered design. The Ardun 1099 intake pushrods are 8.570 inches long overall, 0.385 inch in diameter (at the center), and weigh 71 grams each. The Ardun 1100 exhaust pushrods are 10.00 inches long overall, 0.385 inch in diameter (at the center), and weigh 81 grams each. The Ardun 1025 ball joint was attached to the camshaft lifter end of the pushrod, and the Ardun 1026 ball socket was attached to the rocker arm end of the pushrod, creating a one-piece unit. The pushrods used with the first-design Ardun camshaft lifters are estimated to be 1.50 inches longer than the pushrods used with the second-design Ardun camshaft lifters. The stock Ardun pushrods are solid and heavy; as a result, they will not be used in the Ardun engine described here.

This set of nonadjustable 5/16-inch-diameter Smith Brothers Pushrods was installed in the Ardun engine. The shorter intake pushrods are 8.432 inches long and weigh 50 grams each. The longer exhaust pushrods are 9.830 inches long and weigh 55 grams each.

Smith Brothers Pushrods (see Appendix B: Resources) has been in business since 1953, and the company produces high-quality pushrods at a reasonable price. We installed a set of Smith Brothers Pushrods heat-treated 4130 chrome-moly steel nonadjustable intake pushrods (5/16-inch [0.3125-inch] diameter, 8.432 inches long overall, each weighing 50 grams) and heat-treated 4130 chrome-moly steel nonadjustable exhaust pushrods (5/16-inch [0.3125-inch] diameter, 9.830 inches long overall, each weighing 55 grams). The current retail price for a set of these Smith Brothers Pushrods is $190.

Prior to the final engine assembly, mock up the engine without installing the connecting rod and piston assemblies. Set the correct valve lash. Check the clearance between the pushrods and pushrod holes in the cylinder heads to ensure that the pushrods do not rub against the cylinder heads or valve guide bushing bosses (previously mentioned). Rotate the crankshaft through 360 degrees for each cylinder and check the pushrod clearance for each cylinder. If the pushrod holes in the cylinder heads must be enlarged, make certain to hire a machine shop known for its expertise with Ardun cylinder heads. Otherwise, the chance of striking water will become a certainty.

The first-design Ardun rocker arm assemblies did not use Ardun 1006 rocker arm shaft outer end supports, nor did they use rocker arm locating springs, Ardun 1076 rocker arm shaft retaining rings (circlips), or Ardun 1053 end washers; Ardun 1077 washers were used instead. Each Ardun first-design rocker arm assembly consists of the following parts:

Two each: Ardun 1006 intake rocker arm shaft end supports
One each: Ardun 1003 intake rocker arm shaft center support
One each: Ardun 1021 intake rocker arm shaft, 5/8 inch o.d.
Three each: Ardun 1079 rocker arm shaft support Allen head set screws
Three each: Ardun 1078 jam nuts
Eight each: Ardun 1065 rocker arm bushings
Two each: Ardun 1008L left-hand intake rocker arms
Two each: Ardun 1008R right-hand intake rocker arms
Eight each: Ardun 1053 rocker arm shaft end washers
Eight each: Ardun 1077 rocker arm shaft washers
Eight each: Ardun 1076 rocker arm shaft retaining rings
Eight each: Ardun 1053 end washers
Eight each: Ardun 1077 washers
Eight each: Ardun 1027 rocker arm adjustment screws
Eight each: Ardun 1075 jam nuts
Four each: Allen head end plugs
Two each: Ardun 1005 exhaust rocker arm shaft supports
One each: Ardun 1020 exhaust rocker arm shaft, 5/8-inch o.d.
Two each: Ardun 1007L left-hand exhaust rocker arms
Two each: Ardun 1007R right-hand exhaust rocker arms

The lack of rocker arm shaft outer end supports would have permitted the rocker arm shafts to flex, creating a very undesirable situation. The second-design rocker arm assemblies are the superior model. In order to overcome

This pair of first-design Ardun rocker arm assemblies was fitted with Hawaii Motor Rebuilders rocker arm shaft outer end supports.

the problem of rocker arm shaft flex, Hawaii Motor Rebuilders fabricated a nice set of steel rocker arm shaft outer supports for the first-design Ardun rocker arm assemblies. These are quality items, considering they were fabricated in 1955 without the aid of CNC technology.

The design of the Ardun shaft-mounted rocker arm assembly was innovative in 1947, considering that even today most OHV engines are fitted with stud-mounted rocker arms. If you are not experienced with disassembly and reassembly of Ardun rocker arm assemblies, do not touch them. There are very few people in the world today who have the knowledge and experience to service Ardun rocker arm assemblies. Find an experienced machine shop or a person known for their expertise with Ardun engines to do the job. If you insist on cleaning and reassembling the Ardun rocker arm assemblies yourself, take some good photographs or make accurate sketches as they are disassembled. Take your time and do the job properly. Remember, the parts that make up the Ardun rocker arm assemblies are almost impossible to replace.

The original Ardun rocker arms were manufactured using high-tensile nickel-chromium molybdenum steel, and the rocker arm bushings were lubricated by a full-pressure oiling system. The rocker arm ratio is 1.20. The camshaft lobe lift, less the valve lash, is multiplied by 1.2, resulting in the net valve lift. Ardun Enterprises manufactures a superb set of billet aluminum roller tip rocker arms with a 1.375 ratio. The roller tips are 17-4 heat-treated stainless steel, and these rocker arms have an internal oil passage that eliminates the need for rocker arm spray bars. The current price for a set of the Ardun Enterprises rocker arms is $1,600.

Have the rocker arm shafts gently polished on a lathe using 600-grit wet sandpaper and then finish-polished with a pad of Scotch Brite. The rocker arms and cast-aluminum rocker arm shaft stands can be glass beaded, provided a fine grade of glass material is used. Remember to plug the rocker arm bushings in the rocker arms with a plastic tube or metal shaft before glass beading them. The pallets (tips) of the rocker arms that contact the valves should be refaced. This procedure should only be done by a premium-

Ardun Enterprises manufactures billet-aluminum Ardun rocker arm assemblies with a 1.375 ratio. The rocker arm shafts are hard-chromed, and the roller tips are 17-4 heat-treated stainless steel.

quality machine shop in order to avoid completely destroying the rocker arms.

Remove the socket head set screws in the ends of the rocker arm shafts, and thoroughly clean the internal area of the shafts. Make certain all the oil supply holes in the rocker arm shafts for the rocker arm bushings are clear. The rocker-arm-shaft-to-rocker-arm-bushing clearance in a street-performance engine should be 0.0015 to 0.002 inch when using today's unleaded gasoline. Any clearance over 0.002 inch will require the installation of new rocker arm bushings. After all the rocker arm parts are better than "hospital clean," carry out the reassembly using generous amounts of motor oil. Install the rocker adjustment screws and jam nuts, rocker arm shaft support bracket set screws and jam nuts, and rocker arm shaft socket head set screws using antiseize compound. It is not unusual to find the rocker arm adjusting screw jam nuts and rocker arm shaft support bracket set screws and jam nuts rusted or damaged. These items can be replaced using stainless-steel components.

The rocker arm assemblies used in the Ardun engine described in this book were in excellent condition. The rocker arm assemblies were completely dismantled and thoroughly cleaned, and the rocker arm shafts were polished. The rocker arm pallets (tips) did not require refacing. After glass beading, the rocker arms and rocker arm shaft outer end supports were painted with Glyptal G-1228A medium grey gloss enamel.

During the trial assembly of the blown Ardun engine, it was apparent that the rocker arm tips were not centered over the valve stem tips. The rocker arms could not move over, because they were already against the rocker arm shaft supports. Luke's Custom Machine & Design milled the required amount of material from the sides of the rocker arms to correctly center the rocker arms. We upgraded the Ardun first-design rocker arm assemblies with Ardun second-design rocker arm locating springs, which are no longer available. A local spring shop fabricated the locating springs with 0.048-inch-diameter spring wire. Two rocker arm assemblies require eight pieces of 1/4-inch-long locating springs and eight pieces of 2-inch-long locating springs.

Water Outlets

The first-design Ardun cylinder heads were fitted with an Ardun 1018 aluminum water offtake (water outlet elbow), which was secured in position using an Ardun 1012 gasket, Ardun 1033 stud, Ardun 1034 washer, and Ardun 1035 acorn nut. The Ardun 1019 aluminum blanking cover (rear water outlet elbow cover) was secured in position using an Ardun 1012 gasket, Ardun 1032 stud, Ardun 1034 washer, and Ardun 1035 acorn nut. As previously mentioned, the second-design Ardun cylinder heads were fitted with an Ardun 3402 aluminum water outlet elbow, located at the top between the two center intake ports. This water outlet elbow was secured in position using three studs, flat washers, and acorn nuts. None of the Ardun water outlet elbows or blanking cover gaskets are available today, which means you will have to fabricate them from a sheet of good-quality gasket material. Install the water outlet elbow studs using antiseize compound, and buff the acorn nuts for that uptown look.

Using the first-design Ardun cylinder heads with an Ardun 1018 water outlet elbow will create a clearance problem when a 1949–1953 Ford distributor is installed. There is no room for the Ardun 1018 water outlet elbow with that style of distributor. Luke's Custom Machine & Design fabricates a flat aluminum water outlet elbow that bolts onto the front of the Ardun cylinder head, and an angled smaller diameter water outlet elbow screws into the side of the water outlet elbow. Install Dynaline Industries 54135 neoprene rubber O-rings (2 3/8-inch o.d. by 2 3/16-inch i.d.) between the water outlet elbow and the cylinder head to ensure a watertight fit. Install this setup in both cylinder heads to give the Ardun engine a symmetrical appearance.

This pair of Ardun 1018 aluminum water outlet elbows was used on the first-design Ardun cylinder heads.

Luke's Custom Machine & Design fabricated these compact aluminum water outlet elbows for the blown Ardun engine.

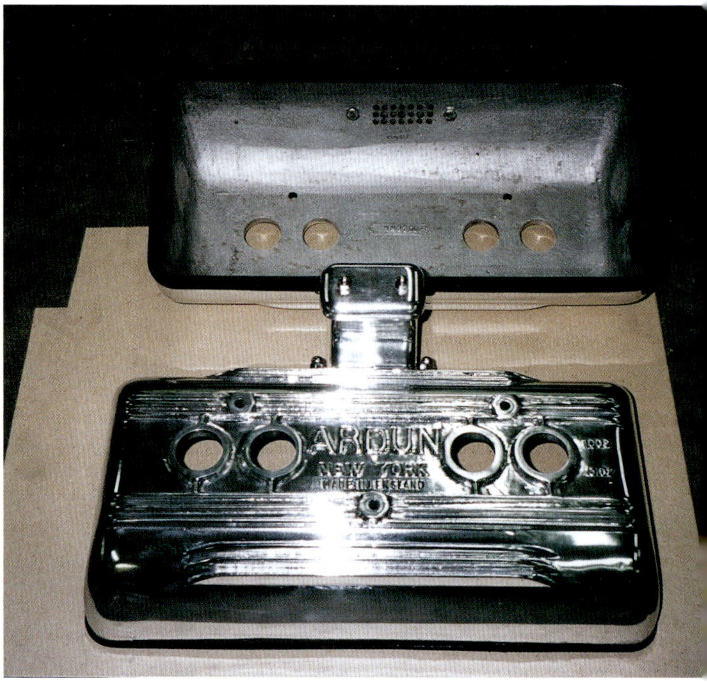

The rare valve covers installed on the blown Ardun engine have "Made in England" cast in the center. Moon Equipment Company 167-ARC vintage valve cover breathers and Luke's Custom Machine & Design neoprene rubber gaskets have been installed.

Although adding Ardun cylinder heads to a Ford flathead V-8 engine block will permit the engine to run cooler, the engine will need thermostats because it will be driven on the street. Removing the thermostats will cause the coolant in the engine to move too fast for the heat to dissipate. The design of the water outlet elbows fabricated by Luke's Custom Machine & Design for the Ardun cylinder heads does not permit the installation of thermostats within their bodies. The outlet diameter is 1 inch, which should restrict the flow of the coolant enough to dissipate the heat.

Valve Covers

The Ardun 1002 valve covers (rocker covers) were cast using nonsonorous (sound-deadening) aluminum. The Ardun valve covers have "Ardun" and "New York" (in smaller letters below "Ardun") cast into the outside center surface. Some of the first (and extremely rare) models also have "Made in England" (in still smaller letters) cast below "New York," while other valve covers have "Ardun" only. Some valve covers have a small brass tag with "Made in England" attached by two rivets below "Ardun." The valve covers manufactured in England for the Allard engine have a large rectangular plate with rounded ends and the words "Allard Motor Company" on them. Each valve cover was secured with three Ardun 1031 flat washers and three Ardun 1015 acorn nuts. In my opinion, it should be mandatory to buff the Ardun valve covers.

Ardun 1016 valve cover gaskets are no longer available. Luke's Custom Machine & Design has Ardun valve cover gaskets manufactured by a gasket company using neoprene rubber; they are channel shaped and reusable, and slip onto the lip of the valve cover. The ends of the valve cover gaskets are placed at the top of the valve cover and glued together. Luke's Custom Machine & Design is the only source of Ardun valve cover gaskets that I am aware of, and the current retail price for a pair is $45.

Ardun 1002 valve covers are not fitted with any breathers or oil filler caps. The single crankcase breather/oil filler tube located at the rear of the valley cover is not sufficient by itself to relieve the crankcase pressure in a blown Ardun engine; as a result, valve cover breathers should be installed. Valve cover breathers were usually installed on the front of the valve covers on the early Chrysler Hemi engines. However, the location of the distributor and generator V-belt would interfere with the valve cover breathers on an Ardun engine if they were mounted in that position. The ideal place to mount valve cover breathers on the Ardun valve covers is at the top in the center. Mooneyes

Luke's Custom Machine & Design built this wooden pattern, showing a cut-away profile of the Ardun cylinder head.

USA (see Appendix B: Resources) manufactures polished-aluminum valve cover breathers with an internal baffle. Mooneyes USA MP1800 right-angle valve cover breathers (3 inches wide) are suitable for the Ardun valve covers.

Estimated Costs

It is almost impossible to estimate a price for a used Ardun conversion kit in good condition; market forces determine what that value would be. In order to arrive at an estimated cost for the complete Ardun conversion kit, I have based the total cost on $10,000. This price might appear high to some readers, although I recently saw a complete second-design Ardun conversion kit (cylinder heads 741 and 748) advertised for $19,500 on eBay. Two flathead engines were included in the deal; however, they appeared as rusty and dirty as the Ardun components. The seller must have been hallucinating, considering a new "state-of-the-art" Ardun conversion kit from Ardun Enterprises is about $13,000.

If the pushrods, valves, valve springs, valve spring seat cups, valve spring retainers, valve stem locks, cylinder head stud kit, and intake manifold are estimated to be $2,500, this would leave $7,500. The pair of cylinder heads, with valve guides, valve seat inserts, and water outlet elbows are estimated to be $5,000. The pair of complete rocker arm assemblies is estimated to be $1,000. The pair of valve covers with spark plug tubes is estimated to be $800. The camshaft lifters are estimated to be $350, and the valley cover is estimated to be $350. These are the values I have used for the blown Ardun engine described here.

Part B

- New Clevite 77 SH-21S camshaft bearings installed allowing for 0.002 inch camshaft clearance and 0.007 inch camshaft end play; number 1 and 2 bearings: SH-21, housing bore: 1.9275 to 1.9285 inches; number 3 bearing: SH-22, housing bore: 1.9275 to 1.9285 inches; camshaft journal diameter: 1.7965 to 1.7970 inches; **$27.95**
- New Crower Cams & Equipment Company Ardun Special blower grind cast-iron alloy solid lifter camshaft with advertised duration: 277 degrees intake and exhaust; duration at 0.050-inch lift: 222 degrees intake and exhaust; lobe separation angle: 112 degrees; net valve lift (with 1.20 ratio rocker arms): 0.386 inch intake and 0.384 inch exhaust; valve lash (hot): 0.010 inch intake and 0.012 inch exhaust; camshaft degreed; 1949–1950 Ford 7RA-6254-B front camshaft gear (camshaft to distributor drive gear) and Ford 6254 rear camshaft gear (camshaft to oil pump idler gear) installed; **$284.96**
- Ardun 1024 high-tensile nickel-chromium molybdenum steel milk-bottle-shaped solid lifters, bases refaced, stock 1.00-inch diameter, 3.260 inches long, installed; weight: 100 grams each; **$439.03 (estimated)**
- New Smith Brothers Pushrods heat-treated 4130 chrome-moly steel nonadjustable intake pushrods (5/16-inch [0.312-inch] diameter, 8.432 inches long overall) installed; weight: 50 grams each; new Smith Brothers Pushrods heat-treated 4130 chrome-moly

steel nonadjustable exhaust pushrods (5/16-inch [0.312-inch] diameter, 9.830 inches long overall) installed; weight: 55 grams each; **$190**

- New Crane Cams 99480-16 aircraft-quality, silicon chrome, single helical coil valve springs with damper; installed height: 90 pounds at 1.578 inches valves closed, and 260 pounds at 1.200 inches valves open (0.400-inch lift); 1.460-inch o.d. and 1.070-inch i.d.; coil bind: 0.935 inch; maximum lift: 0.510 inch; color: red/olive green; new Iskenderian Racing Cams 3607-ST heat-treated chrome-moly steel 7-degree valve spring retainers (3/8-inch stem diameter, 0.060 inch higher than stock) installed with new Crane Cams 99098-1 machined 4140 chrome-moly heat-treated steel 7-degree valve stem locks, standard height, black-oxide finish; Ardun 1071 spring seat cups installed; new Manley Performance Parts 42108-16 heat-treated chrome-moly steel lash caps (3/8-inch stem diameter, 0.220-inch minimum tip thickness) installed; **$231.61**
- Ardun 1062 intake valves reduced to 1 3/4-inch (1.750-inch) head diameter, 3/8-inch (0.375-inch) stem diameter, 4.860-inch overall length; weight: 114 grams each; Ardun 1063 austenitic exhaust valves (1 1/2-inch [1.50-inch] head diameter, 3/8-inch [0.375-inch] stem diameter, 4.865-inch overall length); weight: 101 grams each; valves installed with Serdi-machined multiangle blueprint valve grind; new Enginetech S2890 viton positive-seal valve stem seals installed on intake and exhaust valve guides; **$524.67**
- New Melling 2702 helical aluminum camshaft timing gear (44 teeth) installed with Ford 6258 lock washer plate and Ford 350400 camshaft gear bolts using Loctite and torqued to 20 ft-lb; 1951–1953 Ford C8RT-6059-B/COBA-6059-A polished cast-aluminum timing gear cover installed with new polished stainless-steel hex head bolts and lock washers using Loctite and torqued to 20 ft-lb; timing gear cover gasket installed using silicone sealant; **$169.12**
- Ardun 1001 (first-design) overhead valve 355 T-6 polished cast-aluminum hemispherical combustion chamber cylinder heads (89-cc combustion chamber volume); Ardun 1028 phosphor bronze intake valve seat inserts and Ardun 1029 phosphor bronze exhaust valve seat inserts installed; Ardun 1013 phosphor bronze intake valve guides and Ardun 1014 phosphor bronze exhaust valve guides knurled and installed; new K-Line KL1842STA universal-length 3/8-inch (0.375-inch) bronze bullet valve guide liners installed; Luke's Custom Machine & Design new polished aluminum and chrome water outlet elbows installed with new Dynaline Industries 54140 neoprene rubber O-rings (2 3/8-inch o.d. by 2 3/16-inch i.d.) and new polished stainless-steel countersunk socket head bolts using antiseize compound; cylinder head identity numbers: 309 and 310; cylinder head weight (bare): 30 pounds each; **$6,417.83 (estimated)**
- Cylinder heads pressure tested, glass beaded, and surfaced; Champion J-202 spark plug adapters removed; new Victor/Reinz GX-13173-1 nitroseal cylinder head gaskets installed; new Ardun Enterprises (ARP) cylinder head stud kit (190,000 psi) installed using Permatex aviation form-a-gasket with parallel-ground hardened flat washers and 12-point cylinder head stud nuts installed using molykote and torqued to 50 ft-lb; new Dynaline Industries 54012 neoprene rubber O-rings (1/2-inch o.d. by 3/8-inch i.d.) installed under cylinder head stud flat washers; intake-valve-to-piston clearance: 0.195 inch; exhaust-valve-to-piston clearance: 0.205 inch (with cylinder head gasket); **$1,437.54**
- Ardun first-design complete rocker arm assemblies with Ardun 1021 and 1020 rocker arm shafts; Ardun 1008L and 1008R high-tensile nickel-chromium molybdenum steel intake rocker arms; Ardun 1007L and 1007R high-tensile nickel-chromium molybdenum steel exhaust rocker arms; Hawaii Motor Rebuilders steel rocker arm shaft outer end supports and new rocker arm locating springs installed; new stainless-steel rocker arm adjusting screw jam nuts and rocker arm shaft support bracket set screws and jam nuts installed using antiseize compound; rocker arms and rocker arm shaft outer end supports painted with Glyptal G-1228A medium grey gloss enamel; rocker arm assemblies rebuilt by Luke's Custom Machine & Design; **$1,894.90 (estimated)**
- Ardun 1002 polished nonsonorous aluminum valve covers installed with Ardun 1070 valve cover studs, Ardun 1031 flat washers, and Ardun 1015 acorn nuts using antiseize compound and torqued to 15 ft-lb; Luke's Custom Machine & Design new neoprene rubber valve cover gaskets installed; Moon Equipment Company 167-ARC polished-aluminum right-angle valve cover breathers (3 1/2 inches wide, internally baffled) installed with new polished stainless-steel hex head bolts, AN flat washers, and acorn nuts using antiseize compound; **$1,040.74 (estimated)**

Camshaft and Cylinder Head Total: $12,658.35

CHAPTER 26
INTAKE SYSTEM

Part A
Intake Manifold

The Ardun intake manifold consists of two Ardun 1009 cast-aluminum logs, one bolted to each Ardun cylinder head, along with an Ardun 1069 connector joining the two logs. A single Holley 94 or Stromberg 97 carburetor was mounted on top of the Ardun 1069 connector (or, as it was described by the Ardun Engine Company, the carburetor adapter). To put it politely, this intake manifold was not the greatest of designs.

The Ardun Engine Company also offered the "special racing intake manifold," which had two Ardun 1069 connectors, one at each end of the intake manifold, permitting the installation of two Holley 94 or Stromberg 97 carburetors. It is highly doubtful that any NHRA records were ever set with this intake system!

Ken Austin (see Appendix B: Resources) has been making custom-cast intake manifolds in the United States since 1950 and Ardun intake manifolds since 1997. His Ardun intake manifolds are extremely well-designed, beautiful pieces of work, and they are likely the finest intake manifolds available for the Ardun engine today. Austin intake manifolds have a unique floating flange sealed by a 3/16-inch cross-section O-ring, no welding is required, and they are guaranteed to fit regardless of age or modification to the Ardun cylinder heads. O-rings eliminate the need for conventional intake manifold gaskets. The unpolished Ardun intake manifolds range in price from $850 to $1,250. The Ardun intake manifolds available are:

Model DD: two deuces, dual plane, $1,250
Model DS: two deuces, single plane, $850
Model TD: three deuces, dual plane, $1,150
Model TS: three deuces, single plane, $900
Model FS: four deuces, single plane, $950
Model QD: four-barrel, dual plane, $1,250
Model QS: four-barrel, single plane, $850
Model BB: blower base, $750
Model UB: universal base, $750
Universal base available with different top plates: $150

We installed an Austin Model BB blower intake manifold on the blown Ardun engine described in this section. Although it is unlikely, any casting flash in the intake runners should be ground off. The blower mounting plate

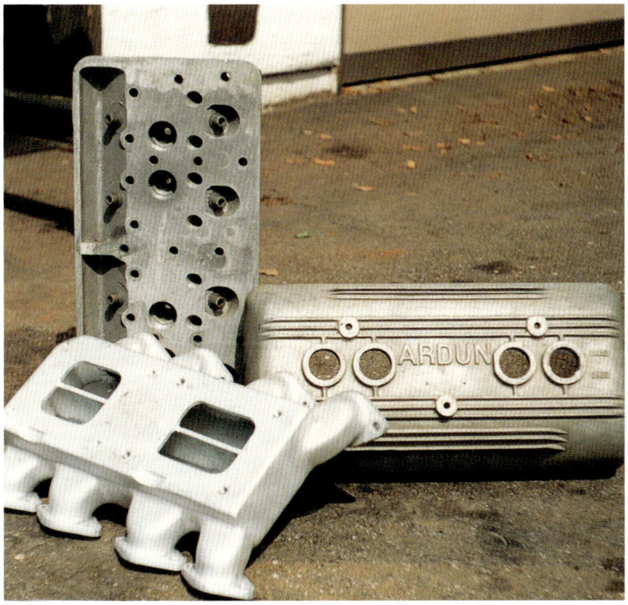

Luke's Custom Machine & Design fabricated this Ardun two four-barrel aluminum intake manifold for Doug King (see Appendix B: Resources). The intake runners are from Ardun 1009 intake manifold logs.

The top of the Austin Model BB blower intake manifold after the plenum opening for a GMC 4-71 blower has been provided.

on the top of the Austin Model BB does not have plenum openings, and there are no stud holes for securing the blower. These openings will have to be added, which will permit the blower to be positioned in the best possible location on the intake manifold.

Pop-off valves are not provided with the Ardun Model BB blower intake manifold. If a GMC 4-71 blower is used, there is no room at the front or rear of the top of the intake manifold to install pop-off valves. Luke's Custom Machine & Design fabricates a screw-in pop-off valve using a small-block Ford 351C (Cleveland) 2.19-inch-diameter intake valve. The valves even have the Ford script on them for a nice detailing touch. One pop-off valve screws horizontally into the right front corner (passenger's side) of the intake manifold, and the second pop-off valve screws horizontally into the left rear corner (driver's side) of the intake manifold.

Luke's Custom Machine & Design fabricated these pop-off valves for the Austin Model BB blower intake manifold. The intake valve is 2.19 inches in diameter, designed for a small-block Ford 351C engine.

In order to install an Austin intake manifold on an Ardun engine, slide the flanges over the intake runners with the beveled side facing outward. Four of the flanges are machined on the outside face. These are for the left (driver's) side of the engine block where the cylinder head sits slightly lower on the engine block than the cylinder head on the right (passenger's) side. Slide the O-rings over the intake runners against the flanges. Install the intake manifold studs in the left (driver's) side cylinder head only. Then place the intake manifold in position and install the intake manifold studs for the other cylinder head. The intake runners on the left (passenger's) side of the intake manifold are slightly longer than the intake runners on the right (driver's) side. The Austin intake manifold will be level when correctly installed.

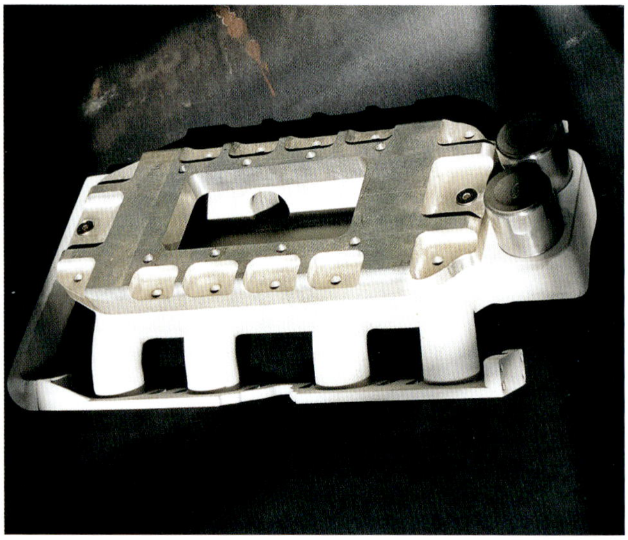

This is the first Ardun blower intake manifold ever fabricated by Luke's Custom Machine & Design. The pop-off valves are located at the rear of the intake manifold.

The Ardun intake manifold studs might be missing or damaged. These can be replaced using Papco 264-048 engine studs (3/8 inch by 2 inches long). The studs can be reduced in length to accommodate any intake manifold. They have 3/8-inch-N.C. threads on the cylinder head end and 3/8-inch-N.F. threads on the intake manifold end. In order to look as elegant as the intake manifold, purchase 16 stainless-steel 3/8-inch-N.F. acorn nuts and 16 AN flat washers. Install the intake manifold studs using antiseize compound, and torque the acorn nuts to 25 ft-lb. Austin intake manifolds buff up beautifully when an eager beaver like "Tim the polishing guy" massages them. He even buffs the underside of the Ardun intake manifolds, because the reflection will be seen off a polished or chrome Ardun 3261 valley cover. If stock intake manifold gaskets are required (when not using an Austin intake manifold), the spacing between the intake port studs (for each intake runner) in the Ardun cylinder head is 3.00 inches, and the intake port diameter is 2.00 inches. The larger exhaust manifold gaskets (tan-colored) that are part of the Fel-Pro FS7525B complete gasket set for the 1949–1953 Ford flathead V-8 engine are perfect candidates.

Luke's Custom Machine & Design fabricated this Ardun blower intake manifold for a customer in Tarzana, California. The pop-off valves are located on the front of the intake manifold.

Although unrelated to the Ardun engine, Ken Austin also casts intake manifolds for 1949–1951 Lincoln flathead V-8 engines. A blower intake manifold and two types of triple-carburetor intake manifolds are available. Finned aluminum cylinder heads for those engines have just been released as well.

Use a 1939–1948 Ford 9415 crankcase breather/fuel pump stand housing with the Ardun engine. There is no provision for a road-draft/crankcase breather tube or a crankcase breather/oil filler tube at the front of the Ardun 3261 valley cover.

GMC 4-71 Blower

Refer to Chapter 16 for information about the GMC 4-71 blower. The Austin Model BB blower intake manifold is designed to accept a GMC 4-71 or GMC 6-71 blower case. The GMC 4-71 blower case modified by Luke's Custom Machine & Design does not have any mounting flanges. As a result, eight holes are drilled through the Austin Model BB blower intake manifold for studs used to secure the blower case to the intake manifold. ARP 200-2404 carburetor studs (5/16-inch N.C. by 3.70 inches long overall) are installed in the bottom of the blower case. The blower case is lowered on the intake manifold, and neoprene rubber O-rings, stainless-steel flat washers, and nylocks are attached to the underside of the intake manifold. The intake manifold and blower mount on the engine as a one-piece unit.

Luke's Custom Machine & Design machined this GMC 4-71 blower case on a milling machine, cut off the mounting flanges, and shaped the sides.

Luke's Custom Machine & Design built this Ardun intake manifold, blower setup, and air cleaner for a customer in Houston, Texas. The blower is an extremely rare Italmeccanica/S.CO.T. model, which looks very classy when polished.

This rear cover began life as a GM 5120726 front cover for double reduction gears on a Detroit diesel engine. Refer to Chapter 16 for a picture of the original cover.

Blower Snout

Refer to Section II

Idler Pulley and Bracket

Refer to Section II

Luke's Custom Machine & Design completed this GMC 4-71 blower with a sculptured snout and ARP 200-2404 studs attached. It is now ready to mount on the Austin Model BB blower intake manifold.

Blower Drive Pulleys

We installed a Blower Drive Service 6330 aluminum crankshaft blower drive pulley (30 teeth, 3 1/2 inches wide, and 1/2-inch pitch [Gilmer]) and a Blower Drive Service 6332 aluminum snout drive pulley (32 teeth, 3 1/2 inches wide, and 1/2-inch pitch [Gilmer]). We also installed a Blower Drive Service 540H200 blower drive belt (1/2-inch pitch [Gilmer], 2 inches wide, and 54 inches

The brand-new Holley 0-4776C classic-look 600-cfm double-pumper four-barrel carburetor provides instant throttle response.

long). The blower drive pulleys are reduced to 2 1/2 inches wide on a lathe. The above combination of blower drive pulleys will provide the Ardun engine with a blower drive ratio of approximately 6.20 percent underdrive and a blower boost of approximately 7.25 pounds. This is a sensible amount of boost for a blown street-performance Ardun engine. Also refer to Chapter 16.

Carburetion

We installed a Holley 0-4776C classic-look 600-cfm double pumper four-barrel carburetor on the blown Ardun engine. This model of carburetor looks entirely suitable on top of a GMC 4-71 blower and is the smallest double pumper carburetor that Holley provides. Also refer to Chapter 16.

Fuel-Pressure Regulator and Fuel Line

Refer to Section II

Air Cleaners

Refer to Section I

Valley Cover

The Ardun 3261 cast-aluminum valley cover is used to cover the intake ports and valve lifter gallery of the Ford flathead V-8 engine, and it fits in place of the stock flathead intake manifold. The stock flathead intake manifold gasket is used with the valley cover, and eight Ford 20448-S intake manifold bolts secure it in position. Cover the intake manifold bolts with chrome acorn nut covers. Trace the outline of the valley cover on the intake manifold gasket and trim it accordingly. The Ardun 3261 valley cover is fitted with a stock position generator/fan mount on the front. The valley cover will look exceptional when buffed or chrome plated.

In order to clear the Austin Model BB blower intake manifold, the front of the crankcase breather/fuel pump stand boss and the top of the three reinforcing fins for the generator mount on the Ardun 3261 valley cover had to be milled slightly.

Part B

- Ardun 3261 polished-aluminum valley cover installed with Ford 20448-S intake manifold bolts and new stainless-steel AN flat washers using Loctite and torqued to 25 ft-lb; valley cover gasket installed using silicone sealant; new Speedway Motors 910-10125 chrome acorn nut covers for 9/16-inch head installed; new Jitney Auto Parts polished-aluminum crankcase breather/fuel pump stand housing and new Mr. Gasket 2060 chrome push-in breather cap installed with new polished stainless-steel acorn nuts and lock washers using antiseize compound; crankcase breather/fuel pump stand housing gasket installed using silicone sealant; **$555.44 (estimated)**

"Tim the polishing guy" buffed this Ardun 3261 aluminum valley cover and now it looks outstanding.

- New Austin Model BB polished-aluminum blower intake manifold installed with new Papco 264-048 engine studs (3/8 inch by 2 inches long, reduced to 1 3/4 inches) and new polished stainless-steel acorn nuts and AN flat washers using antiseize compound and torqued to 25 ft-lb; Luke's Custom Machine & Design new front and rear pop-off valves fabricated with Ford OHV 2.19-inch-diameter intake valves; **$1,663.90**
- GM 5138516 polished-aluminum small-bore die-cast Model 4-71 (blower) supercharger case; GM 5122363 polished die-cast aluminum front and rear bearing plates; new Blower Drive Service 671SL1 bearing seals; new MRC 5205 front bearings; new MRC 6205 rear bearings; and GM 5114442 polished die-cast aluminum front cover assembled with new polished stainless-steel socket head bolts, lock washers, and AN flat washers using antiseize compound; new Blower Drive Service 471GK gasket kit installed using Permatex aviation form-a-gasket; new Blower Drive Service 556 oil-level sight gauge and new stainless-steel oil filler and oil drain plugs (3/8-inch NPT) installed using pipe thread sealant; pressure-relief system fabricated using new Weatherhead 69-3 polished-brass male-elbow compression fitting (3/16-inch tube to 1/8-inch NPT); new Weatherhead 68-3 polished-brass male-connector compression fitting (3/16-inch tube to 1/8-inch NPT); new 3/16-inch o.d. polished-copper coiled tubing; blower case pruned and shaped; rotors double-pinned and clearanced; front and rear bearing plates machined for bearing support rings; GM 5120726 polished-aluminum rear cover fabricated by Luke's Custom Machine & Design; blower installed with new ARP 200-2404 studs (5/16-inch N.C. by 3.70 inches long overall); new neoprene rubber O-rings; new stainless-steel flat washers and nylocks using pipe thread sealant; blower gears lubricated with Castrol Syntec 75W-90 gear oil; **$1,830.28**
- Luke's Custom Machine & Design new LS2000 polished 6061 T-6 aluminum sculptured blower drive snout; new LS2200 blower drive snout shaft, 4140 chrome-moly heat-treated steel; new LS4100 blower drive snout coupler, 4140 chrome-moly heat-treated steel, installed with new stainless-steel socket head bolts using antiseize compound; **$500**
- Luke's Custom Machine & Design new polished 6061 T-6 aluminum blower belt idler pulley (5-inch o.d.) and new polished-aluminum idler pulley adjustment bracket installed with new polished stainless-steel AN flat washer, lock washer, and nut using antiseize compound; **$629.63**
- New Blower Drive Service 6330 polished-aluminum crankshaft blower drive pulley (30 teeth, 3 1/2 inches wide, and 1/2-inch pitch [Gilmer]) and new Blower Drive Service 6332 blower snout drive pulley (32 teeth, 3 1/2 inches wide, and 1/2-inch pitch [Gilmer]) installed with new polished stainless-steel socket head bolts, lock washers, and AN flat washers using antiseize compound and torqued to 25 ft-lb; blower drive pulleys reduced to 2 1/2 inches wide; new Blower Drive Service 540H200 blower drive belt (1/2-inch pitch [Gilmer], 2 inches wide, and 54 inches long) installed; blower drive ratio: 6.20 percent underdrive; boost: 7.25 pounds (estimated); **$295**
- New Holley 0-4776C classic-look nonemission 600-cfm four-barrel double-pumper carburetor installed with new ARP 400-2401 polished stainless-steel carburetor studs (standard length) using antiseize compound; new Blower Drive Service 14B4 polished-aluminum carburetor-to-blower adapter (1 inch thick) installed with new stainless-steel socket head bolts using antiseize compound; new Super Sunny S6052 chrome carburetor return spring installed; **$527.33**
- New Mr. Gasket 1552 chrome dual-inlet fuel line (9 5/16 inches center to center) installed; new Holley 12-803 fuel-pressure regulator (adjustable from 4 1/2 inches to 9 psi) installed with new Weatherhead 1069-6 polished-brass 90-degree male elbow (3/8-inch-i.d. hose to 1/4-inch NPT); new Weatherhead 3320-6-4 polished-brass bushing (3/8-inch NPT to 1/4-inch NPT [inlet line]); new Weatherhead 3220-6-2 polished-brass bushing (3/8-inch NPT to 1/8-inch NPT); new Weatherhead 3325-2 brass hex nipple (1/8-inch NPT [outlet line]); new Weatherhead 3220-6-2 polished-brass bushing (3/8-inch NPT to 1/8-inch NPT) and new Holley 26-500 fuel-pressure gauge (1 to 15 psi, 1 1/2-inch-diameter face); all fittings installed using pipe thread sealant; new Auto-Meter 4401 Pro Comp Ultra-Lite vacuum/boost gauge (0 to 30 hg./0 to 20 psi, 2 5/8-inch-diameter face) installed with new Weatherhead 1066-4 polished-brass female connector (1/4-inch i.d. to 1/8-inch NPT); **$122.85**

Intake System Total: $6,124.43

CHAPTER 27
IGNITION SYSTEM

Part A
Distributor

Because building a blown street-performance Ardun engine is an expensive endeavor, you should use an electronic ignition to protect that investment. The distributor used in the Ardun engine described in this section was mounted in the stock 1949–1953 Ford flathead V-8 position. We installed a Mallory 5027501 magnetic breakerless distributor. Refer also to Chapter 17.

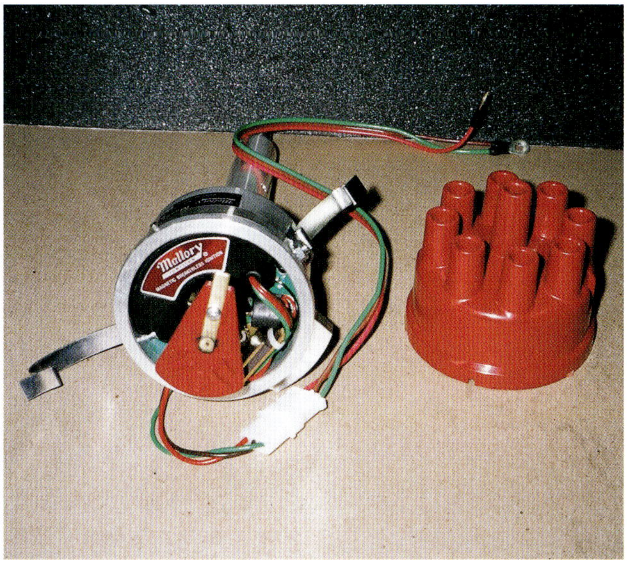

This Mallory 5027501 magnetic breakerless distributor is designed for the 1949–1953 Ford flathead V-8 engine.

The 1949–1953 Ford flathead V-8 distributor was secured in position using the Ford 12270 distributor hold-down clamp that was bolted to the right (passenger's) side cylinder head. There is no allowance on the Ardun 1001 cylinder heads to secure the distributor. As a result, Luke's Custom Machine & Design fabricates a polished stainless-steel bracket that clamps around the distributor neck and bolts to the top of the timing gear cover.

A few brief comments about ignition advance are in order. The fuel mixture in a cylinder must be ignited before the piston reaches the top of the compression stroke, placing the maximum pressure on the piston top for the power stroke. In order to do this, the timing must be controlled so there is more ignition advance at higher engine speeds than there is at lower engine speeds. The ignition advance increases as the engine accelerates. Any vacuum advance is only a consideration at idle or at low cruising speeds where the engine is not laboring.

A rare vintage Scintilla-Vertex magneto with a Vertex angle drive adapter is mounted on the front of this blown Ardun engine.

The crankshaft makes two full revolutions for every single camshaft revolution (the Ford flathead V-8 crankshaft gear has 22 teeth and the camshaft gear has 44 teeth). If the mechanical advance in the distributor is 13 degrees, the actual advance is 26 degrees. If the crankshaft has 10 degrees of advance and the distributor has 26 degrees of actual advance, the result is 36 degrees of total advance. The basic guideline for ignition timing in a blown street-performance engine is between 28 and 36 degrees of total ignition advance. Every engine is different and, therefore, tuned differently.

Coil and Ignition Control

Refer to Section I

Spark Plug Wires and Plugs

Mallory (Mr. Gasket Company) and Taylor/Vertex advise that copper wire core or stainless-steel wire core spark plug wires should not be used with electronic ignitions. In addition to this advisory, the engines described in this book are intended for street use, and there is no valid reason to use copper or stainless-steel wire core spark plug wires for street applications.

The Mopar Performance (Chrysler Corporation) 4120808 Hemi spark plug tube and insulator kit is a decent and practical addition to any Ardun engine. The original Mopar Performance 4120808 Hemi spark plug tube and insulator kit consisted of a spark plug wire/tube cover, a plastic insulator, and a rubber spark plug boot. The new Mopar Performance 4120808 kits do not include the rubber spark plug boot. Molded from the original Chrysler tooling, the Taylor/Vertex 44500 rubber spark plug boot, the 44000 plastic insulator, and the 44300 spark plug wire/tube cover are available individually.

None of the major ignition product manufacturers offer a spark plug wire kit for the Ardun engine, which means a spark plug wire set will have to be compiled. The Taylor/Vertex 35071 Pro Series 8-millimeter high-performance black spark plug wire with Spiro-Pro Kelvar resistor core is available in 30-foot-long packages. This is almost enough spark plug wire to complete two engines.

Mopar Performance (Chrysler Corporation) 4120808 Hemi spark plug tube and insulator kits can be used with Ardun engines.

The Taylor/Vertex 35071 spark plug wire is a top-quality product designed for street use. The Standard ST113 spark plug terminals, Standard ST26 distributor cap terminals, and Standard TN9 distributor cap nipples are available to complete the spark plug wire set. Install a set of Spectre 4245 chrome/plastic spark plug wire separators to give the Ardun engine a neat detailing touch.

I provided comprehensive information about spark plugs for the Ardun engine in the first flathead book and do not intend to repeat all of it here. The Ardun engine described in this book was equipped with Autolite 386 regular-gap 18-millimeter spark plugs. Because the engine utilizes an ignition control box, the recommended spark plug gap is 0.050 inch. Always coat the threads of the spark plugs with antiseize compound when installing them in aluminum cylinder heads to prevent the spark plug threads from galling.

Autolite 386 regular gap 18-millimeter spark plugs were installed in the blown Ardun engine.

Generator

The owner of the Ardun engine described in this section did not want the new look of an alternator on his engine. Instead, a 1956 Ford V-8 generator (rated at 30 amps and 15 volts) was chrome plated, rebuilt, and mounted on the Ardun engine using a polished-aluminum bracket and polished stainless-steel brace fabricated by Luke's Custom Machine & Design.

Starting Motor

Refer to Section I

Spark Plug Tubes and Retainers

Ardun 1051 spark plug tubes (funnels) are screwed into the cylinder heads, Ardun 1011 spark plug tubes seals are positioned on the end of the spark plug tubes, and Ardun 1052 spark plug tube retainers are screwed onto the spark plug tubes against the valve covers. The spark plugs are then inserted in the spark plug tubes and torqued to 15 ft-lb using antiseize compound. Ardun 1051 spark plug tubes are a flimsy design manufactured using thin metal. They usually rust, causing corrosion in the cylinder head area around them. This situation makes it almost impossible to remove the spark plug tubes, and they often collapse when a wrench is used in an attempt to dislodge them. The large 1 1/2-inch diameter of the spark plug tubes means they usually have to be dented with a hammer to clear the rocker arms—very high tech.

In the early 1980s, Luke Balogh designed and machined stainless-steel spark plug tubes with serrated-brass or black-anodized nuts. These are a very upscale replacement for the stock Ardun assembly. The center section of the spark plug tube is 1 1/4 inches in diameter to ensure rocker arm clearance. The serrated nuts accommodate a Mopar Performance 4120808 Hemi spark plug insulator and boot assembly. The price for a set of the Luke's Custom Machine & Design spark plug tubes and nuts is $350.

Installing the Luke's Custom Machine & Design spark plug tubes is a simple matter. Install the spark plug tubes in the cylinder heads, secure the valve covers in position, install neoprene rubber O-rings on the serrated nuts, and screw the serrated nuts onto the spark plug tubes against the valve covers. The Hemi spark plug wire/tube covers slide onto the spark plug wires, the plastic spark plug insulators slide onto the spark plug wires, the rubber spark plug boots slide onto the spark plug wires, and the spark plug terminals are crimped on the ends of the spark plug wires.

When using 18-millimeter spark plugs with Luke's Custom Machine & Design spark plug tubes, the spark plugs must be installed first, followed by the spark plug tubes. The 18-millimeter spark plug socket is too large for the inside diameter of the spark plug tubes. When using 14-millimeter spark plugs, the spark plug tubes can be installed first, followed by the spark plugs.

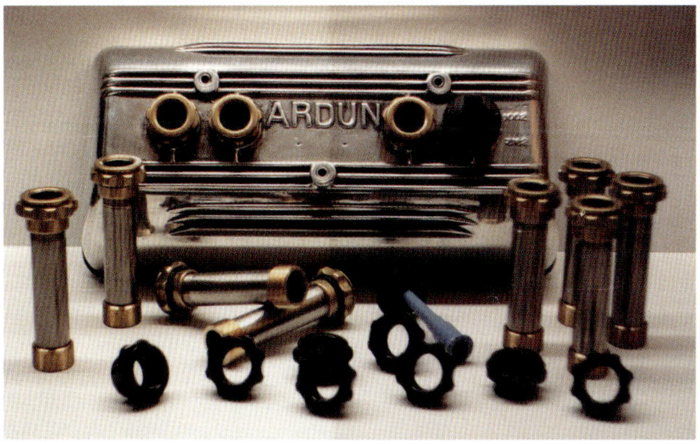

Luke's Custom Machine & Design manufactures the Ardun spark plug tube kits. Pictured here are both the brass and black-anodized serrated nuts.

Part B

- New Mallory 5027501 magnetic breakerless distributor (for 1949–1953 Ford) with full centrifugal (mechanical) advance (6061 T-6 billet aluminum) installed with new Speedway Motors 910-12701 distributor drive gear; Luke's Custom Machine & Design new polished stainless-steel distributor hold-down bracket installed; total distributor advance: 14 degrees at 850 rpm, 22 degrees at 1,800 rpm, and 26 degrees at 2,400 rpm; new MSD 6420 Model 6AL multiple-spark discharge ignition control with rpm-limiter modules and new MSD 8200 Blaster 2 chrome ignition coil (45,000 volts) installed with new polished stainless-steel ignition coil bracket; **$575.52**
- New Taylor/Vertex 35071 Pro Series 8-millimeter high-performance black spark plug wire with Spiro-Pro Kelvar resistor core installed with new Mopar Performance 4120808 Hemi spark plug tube and insulator kits, new Standard ST113 spark plug terminals, new Standard ST26 distributor cap terminals, new Standard TN9 distributor cap nipples, and new Spectre 4245 chrome/plastic professional spark plug wire separators; new Autolite 386 regular-gap 18-millimeter spark plugs installed with 0.050-inch gap using antiseize compound; **$290.66**
- Luke's Custom Machine & Design new stainless-steel spark plug tubes, neoprene rubber O-rings, and new polished-aluminum serrated nuts installed using antiseize compound; **$350**
- 1956 Ford B6A-10002-A chrome generator (rated at 30 amps and 15 volts) and new Super Sunny S8867 polished-aluminum fan installed with Luke's Custom Machine & Design new polished-aluminum generator pulley, new polished-aluminum bracket, and new polished stainless-steel brace using new polished stainless-steel hex head bolts, lock washers, and AN flat washers with antiseize compound; new Koyo 62002RSC3 front bearing and 62032RSC3 rear bearing installed; new Gates 11A1625 high-performance crankshaft/water pumps/generator V-belt (7/16 inch by 64 inches long) installed; **$631.13**
- New Powermaster 9507 XS torque starting motor installed; **$189.95**

Ignition System Total: $2,037.26

CHAPTER 28
COOLING SYSTEM AND MISCELLANEOUS

Part A
Water Pumps

The most common water pumps found on the 1949–1953 Ford flathead V-8 engines are passenger car models. These water pumps have an angled motor mount pad, which is not the greatest design for installation in different frames. The early 1949 Ford 8BA-8501 and 8BA-8502 passenger car water pumps were fitted with a shaft and bushing. The Ford 8BA-8591 water pump rebuild kit is designed for that model. The later 1949–1953 Ford 8BA-8503-D and 8BA-8504-D, or EAB-8501 and EAB-8502, passenger car water pumps were fitted with a shaft and bearing. The Ford A8C-8591 water pump rebuild kit is designed for those models.

The 1949–1951 Mercury flathead V-8 water pumps are similar to the later Ford truck water pump models, except that the horizontal motor mount flange is located higher up on the Mercury water pump body. The 1949 Mercury 8CM-8501 and 8CM-8502 water pumps are fitted with a shaft and bushing. The Ford 8BA-8591 water pump rebuild kit is designed for that model. The 1950–1951 Mercury OCM-8501 and OCM-8502, or OCM-8503 and OCM-8504, water pumps are fitted with a shaft and bearing. The Ford A8C-8591 water pump rebuild kit is designed for those models. The 1952 and 1953 Mercury water pumps are the same as the 1952 and 1953 Ford passenger car water pumps.

The most practical water pumps to use on a 1949–1953 Ford flathead V-8 engine block are the later Ford truck models. These water pumps have a horizontal motor mount flange that makes adapting the flathead engine to most frames a simple procedure. Like many other people, I have usually referred to the later Ford truck

The 1950–1951 Mercury OCM-8503 and OCM-8504 water pumps are mounted on a 1952 Ford flathead V-8 engine block. The aluminum pulley bolted to the front of the 1949–1953 Ford truck GAUB241815 crankshaft pulley is intended for a power steering pump.

The 1948 Ford truck 8RT-8505 and 8RT-8506 water pumps were fitted with a shaft and bushing.

water pumps as the 1949–1953 models. In actual fact, the later Ford truck water pumps were produced from 1948–1952. The 1953 Ford truck water pump is a different design. The early 1948 Ford truck 8RT-8505 and 8RT-8506 water pumps are fitted with a shaft and bushing. There is an oil cup located on the top of the neck of the water pump body used to lubricate the bushing. The Ford 8BA-8591 water pump rebuild kit is designed for that model. The later 1948–1952 Ford truck 8RT-8501-B and 8RT-8502-B, or 8RT-8503-D and 8RT-8504-D, water pumps are fitted with a shaft and bearing. The Ford A8C-8591 water pump rebuild kit is designed for those models. It is common to find aftermarket replacement later-model Ford truck water pumps on flathead engines. The 30115-1 is the left (driver's) side water pump, and the 30114-1, or 1-9601, is the right (passenger's) side water pump.

Complete repair kits (including impellers) for the 1949–1953 Ford flathead V-8 water pumps are not easy to locate, and good used water pumps are becoming harder to find, although Jitney Auto Parts is still a good source for these items. "Speedy" Bill Smith, the founder and president of Speedway Motors, has designed and manufactures new water pumps for 1949–1953 Ford passenger cars and trucks. These new water pumps are a high-nickel casting with a modular impeller and both roller and ball bearings, and they have a 3/4-inch-diameter shaft with a ceramic seal. The Speedway Motors 910-15592L and 910-15592R water pumps are designed for 1949–1953 Ford trucks. The Speedway Motors 910-15595L and 910-15595R are designed for 1949–1953 Ford passenger cars with narrow V-belt pulleys. The Speedway Motors 910-15594L and 910-15594R are designed for 1949–1953 Ford passenger cars with wide V-belt pulleys. The current retail price for each new Speedway Motors water pump is $89.95.

The Ford A8C-8591 water pump rebuild kit (water pump gasket not pictured).

Idler Pulley

The blown Ardun engine described here was installed in a 1940 Mercury coupe. In order for the blower belt to clear the crankshaft/water pumps/generator V-belt and the radiator, the crankshaft/water pumps/generator pulleys had to be located as close as possible to the front of the engine block. As a result, an idler pulley had to be located in such a position as to permit the V-belt to clear the 1949–1953 distributor location. Luke's Custom Machine & Design fabricated an aluminum adjustable idler pulley mounted on the front of the generator mounting bracket. The V-belt now passes under the front of the distributor in an almost horizontal position, and the tension is adjusted with the idler pulley—very clever engineering.

Luke's Custom Machine & Design fabricated this aluminum idler pulley and bracket to permit the V-belt to clear the distributor.

Labor

Costs for the labor for checking clearances, gapping piston rings, degreeing the camshaft, painting and detailing, trial assembly of the motor, final assembly of the motor, blower setup, and the initial startup of the engine are not included in Section B of each chapter in this section of the book. I shall now include that cost here, at the end of the chapter.

The 1948–1952 Ford truck 8RT-8503-D and 8RT-8504-D water pumps and Luke's Custom Machine & Design polished-aluminum water pump pulleys are ready to be installed on the blown Ardun engine.

Part B

- 1948–1952 Ford truck 8RT-8503-D and 8RT-8504-D water pumps rebuilt with NOS (new old stock) Ford A8C-8591 repair kits and installed with new polished stainless-steel bolts and lock washers using Loctite and torqued to 25 ft-lb; water pump gaskets installed using silicone sealant; Luke's Custom Machine & Design new polished-aluminum water pump pulleys installed; new Gates 11A1625 high-performance crankshaft /water pumps/generator V-belt (7/16 inch by 64 inches long) installed; water pumps painted with PPG DGHS 9000 black high-gloss polyurethane; **$401.24**
- Luke's Custom Machine & Design new polished-aluminum crankshaft/water pumps/generator V-belt idler pulley installed with new polished stainless-steel hex head bolt and lock washer using antiseize compound; **$148.15**
- Labor for checking clearances, gapping piston rings, degreeing the camshaft, painting and detailing, trial assembly of motor, final assembly of motor, blower setup, and the initial startup of the engine; **$1,733.33**

Cooling System and Miscellaneous Total: $2,282.72

Note: The estimated output of this engine is 377 horsepower at 6,000 rpm and 375 ft-lb torque at 4,000 rpm (See Appendix A: Dyno Print-Outs).

Blown Ardun Street-Performance Engine Total:

$2,281.87 engine block total
$1,163.57 crankshaft total
$1,718.56 connecting rods and pistons total
$1,362.34 lubrication system total
$12,658.35 camshaft and cylinder heads total
$6,124.43 intake system total
$2,037.26 ignition system total
$2,282.72 cooling system and miscellaneous total
$29,629.10 (U.S.)

CHAPTER 29
MISSING PARTS

Exhaust Manifolds

None of the major exhaust header manufacturers offer exhaust manifolds or headers for an Ardun engine. Custom exhaust headers will have to be fabricated. The spacing between the exhaust port studs in the Ardun cylinder head is 3.00 inches, and the exhaust port diameter is 1 5/8 inches (1.625 inches). Walker 31309 exhaust manifold gaskets are a perfect fit for these cylinder heads.

If the Ardun cylinder head exhaust port studs are damaged or missing, use new Papco 264-048 engine studs. These engine studs have 3/8-inch-N.C. threads on the cylinder head end and 3/8-inch-N.F. threads on the exhaust manifold end and they are 2 inches long. Reduce the engine studs to 1 3/4 inches long by cutting 1/4 inch off the N.F. thread end. Install the engine studs using antiseize compound with new stainless-steel nuts and lock washers, and torque them to 25 ft-lb. The 1 3/4-inch length will accommodate 3/8-inch-thick exhaust header flanges.

This Ford flathead V-8 aluminum flywheel is a piece of real hot rod history. It was manufactured by ALBRO (the forerunner to Scheifer), weighs 12 pounds, is drilled for a 9- or 10-inch-diameter pressure plate, and it has a sprayed brass insert approximately 0.060-inch thick.

Flywheel and Flexplate

One of the easiest later-model transmissions to adapt to a 1949–1953 Ford flathead V-8 engine block is the 1961–1964 Borg-Warner T-10, which was installed in full-size

Here are all the parts required to install a 1961–1964 Borg-Warner T-10 manual four-speed transmission in a 1949–1953 Ford flathead V-8 engine block. The transmission, starting motor plate, and clutch lever have been painted with PPG DGHS 9000 black high-gloss polyurethane.

Luke Balogh stands in his light blue formal evening attire on the left, and I am on the right. The engine on the left is for his 1950 Meteor (Canadian Ford), the blown French flathead is in the center, and the blown Ardun is on the right.

Ford and Mercury passenger cars. This four-speed manual transmission has a nine-bolt curved side cover with a deep notch at the top front of the tailshaft. It is a true bolt-in transmission conversion for any 1949–1953 Ford and Mercury passenger cars, with only one required modification. Reduce the diameter of the front bearing retainer on the Borg-Warner T-10 transmission to 4 5/8 inches in order to fit inside the 1949–1953 Ford 6392 passenger car bell housing.

All the parts required for this conversion are: 1949–1953 Ford 6375 flywheel; Centerforce 280490 clutch disc (10-inch diameter, 1 1/16-inch shaft diameter, and 10 splines [course]); 1949–1953 Ford 7563 long-style pressure plate (10-inch diameter); 1949–1953 Ford 7600 bronze pilot bearing (this can be substituted with a National 203LU roller pilot bearing, 1.575-inch o.d. by 0.665-inch i.d.); 1949–1953 Ford 7580 release bearing with a Ford 7561 hub (throw-out bearing); 1949–1953 Ford 6392 bell housing and Ford 8BA-6366 starting motor plate for a passenger car; 1949–1953 Ford 7515 release lever (clutch fork) with Ford 7605 release lever spring; and GM 14075724 clutch fork rubber boot. There are no other adapters required, with the exception of modifying the vehicle transmission crossmember to accept the Borg-Warner T-10 transmission.

Fan

Refer to Section I

CHAPTER 30
ASSEMBLY

Note: The clearances and torque settings indicated in this book are those I use in my own engines, and I know they work. It is up to the engine builder to ensure the engine is assembled using the proper clearances and torque settings. An accomplished engine builder will follow the clearances specified by the *Ford Motor Company Master Repair Manual* and those recommended by aftermarket equipment manufacturers. Also refer to Sections I and II.

Engine Block

Clean the camshaft bearings with solvent using a clean, lint-free cloth. Install the camshaft bearings dry in the engine block using a camshaft bearing installation tool (this is a must), taking special care to line up the oil holes in the camshaft bearings with the oil holes in the engine block. Do not lubricate the camshaft bearing bosses.

Lubricate the oil pump idler gear bushing and oil pump idler gear shaft with motor oil, and then install the assembly in the rear of the engine block using a rubber mallet to tap the oil pump idler gear shaft in position. The smaller side of the oil pump idler gear faces outward. Install the oil pump idler gear cover gasket using silicone sealant. Install the oil pump idler gear cover and oil pump idler gear cover bolts and lock washers using Loctite and torque them to 15 ft-lb.

Install a new stainless-steel 3/8-inch NPT plug in the oil gallery line at the face of the engine block (behind the camshaft gear) using Permatex aviation form-a-gasket. Install a new stainless-steel 1/8-inch NPT plug in the oil gallery line at the rear of the engine block (above the oil pump idler gear cover) using Permatex aviation form-a-gasket.

Gently tap the aluminum block-off plug into the fuel pump pushrod bushing using Permatex aviation form-a-gasket. Install new stainless-steel 1/4-inch NPT plugs in

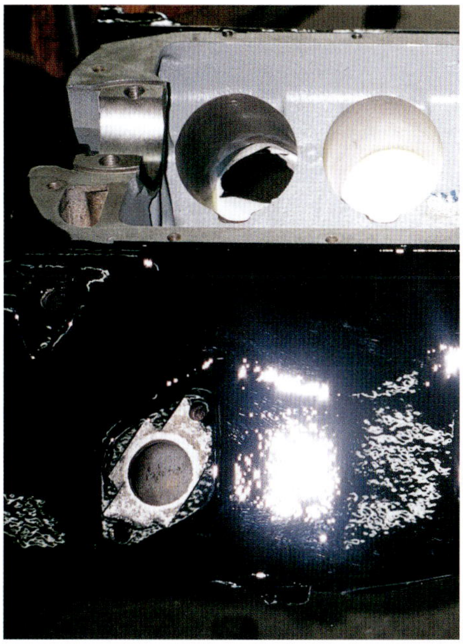

The Ardun engine block was professionally painted with PPG DGHS 9000 black high-gloss polyurethane and is ready for assembly.

Liberally coat the camshaft with Crower Cams & Equipment Company 86093 assembly prelubricant and install, along with the Ford 8BA-6335 upper rear oil seal retainer.

the water jacket drain holes in the front of each side of the engine block (behind the water pumps) using pipe thread sealant. Install a new stainless-steel 1/8-inch NPT plug in the oil filter line hole (near the rear of the left side of the engine block above the oil pan rail) using pipe thread sealant.

Install the four (one on each side of the engine block) stainless-steel 7/16-inch N.C. set screws (used to block off the cylinder head bolt holes) in the engine block using Permatex aviation form-a-gasket. Install a new stainless-steel 1/4-inch NPT plug in the water pump bypass hole at the front of each side of the deck surface using Permatex aviation form-a-gasket.

Install the camshaft gear on the camshaft using Loctite on the bolts, and torque the bolts to 20 ft-lb. Bend the camshaft locking plate tabs over the heads of the camshaft bolts.

Lubricate the camshaft bearings and camshaft bearing journals with Clevite 77 Bearing Guard and generously apply the camshaft assembly lubricant (provided by the camshaft manufacturer) on the lobes of the camshaft. Gently slide the camshaft into the engine block, taking care not to nick the camshaft bearings.

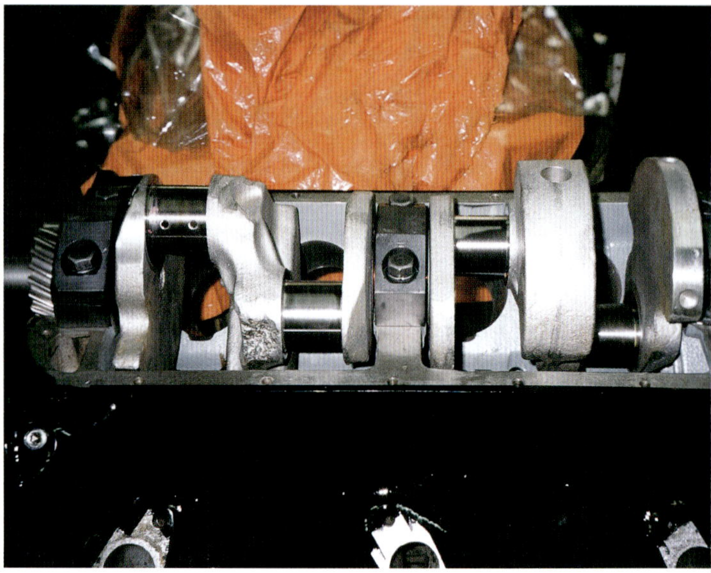

Install the crankshaft and torque the main bearing cap bolts to 105 ft-lb.

Crankshaft

Install the rear main bearing oil seal retainer in the engine block using silicone sealant, and gently tap it into position with a rubber mallet. Install the rear upper cord-material oil seal in the retainer by rolling it into place with a large socket, or use the head of a large ball-peen hammer and tap the head of the hammer with another hammer.

Clean the main bearings with solvent using a clean, lint-free cloth. Line up the locking tang (tab) on the main bearings with the notch in the main bearing saddles, and install the main bearings in the engine block dry. Do not lubricate the main bearing saddles in the engine block. Use Clevite 77 Bearing Guard to lubricate the face of the main bearings after they are installed in the engine block. Line up the locking tang (tab) on the main bearings with the notch in the main bearing caps, and install the main bearings in the main bearing caps dry. Do not lubricate the main bearing caps. Use Clevite 77 Bearing Guard to lubricate the face of the main bearings after they have been installed in the main bearing caps.

Install the stainless-steel connecting rod journal oil line plugs in the crankshaft using Loctite. Place the crankshaft gear Woodruff key in the crankshaft using antiseize compound, and gently tap the Woodruff key into position with a hammer. Apply antiseize compound to the snout of the crankshaft, and slide the crankshaft gear onto the snout as far as it will go by hand. The side with the dot or line on the crankshaft gear faces outward. Line up the crankshaft gear with the crankshaft gear Woodruff key, and gently tap the crankshaft gear into position using a piece of steel pipe a little larger in diameter than the diameter of the crankshaft snout.

Gently lower the crankshaft into the main bearing saddles, taking special care not to nick the crankshaft bearings. Make certain to line up the timing marks on the crankshaft and camshaft gears. The crankshaft gear has a dot or line, and the camshaft gear has a dot or line. These marks should be vertically in line with each other. The main bearing caps are placed in their correct order in the engine block with the locking tang (tab) in the engine block and the locking tang (tab) in the main bearing caps on the same side. Lubricate the main bearing cap bolts with molykote, and torque the main bearing cap bolts to 105 ft-lb. The crankshaft should now turn freely in the engine block. Slide the oil slinger onto the crankshaft snout.

Connecting Rods and Pistons

Remove the wrist pins from the pistons. Coat the wrist pins and the wrist pin bushing in the connecting rods with motor oil. The Ross Racing Pistons forged aluminum blower pistons have a larger notch on the piston top for intake valve clearance and a smaller notch on the piston top for exhaust valve clearance. The larger notch faces upward on the engine block, and the bearing notches in the connecting rods always face outward from the camshaft. Assemble the connecting rods and pistons, and

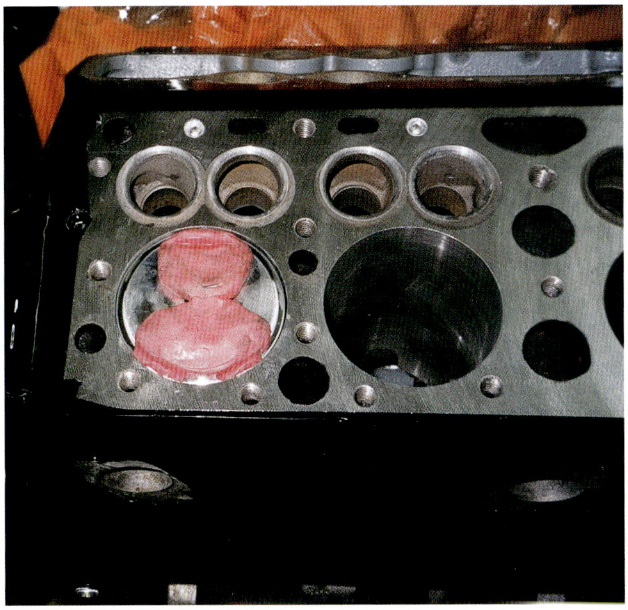

Install one piston and connecting rod assembly with putty on the piston top in order to check the valve-to-piston clearance. The intake valve clearance is 0.195 inch, and the exhaust valve clearance is 0.205 inch with the cylinder head gasket.

install the Spiro-Lox wrist pin retainers by spreading the retainer, inserting one end into the wrist pin hole, and gently but evenly spinning the retainer into the wrist pin hole with a small screwdriver.

Install the piston rings on the pistons using a piston ring expander tool, and rotate each piston ring on the pistons so it is within the manufacturer's recommended arc, also known as "preferred ring gap location."

Clean the connecting rod bearings with solvent using a clean, lint-free cloth. Install the connecting rod bearings dry in the connecting rods and connecting rod caps. Line up the locking tang (tab) on the connecting rod bearing with the notch in the connecting rod and connecting rod cap. Apply Clevite 77 Bearing Guard to the face of the connecting rod bearings.

Apply motor oil to the inside of the piston ring compressor, and place the piston ring compressor around the piston. Tighten the piston ring compressor, insert the connecting rod into the cylinder (with the connecting rod bearing locking tang facing outward from the crankshaft), and gently tap the piston down into the cylinder using a rubber hammer or a piece of wood. Guide the connecting rod around the crankshaft journal as the piston is pushed downward. After the connecting rod is on the connecting rod journal, install the connecting rod cap with the locking tang (tab) on the same side as the locking tang (tab) in the connecting rod. Install the connecting rod bolts using molykote, and torque the connecting rod bolts to 42 ft-lb, allowing for 0.0061-inch stretch. Repeat this procedure for the rest of the connecting rod and piston assemblies.

Install the piston and connecting rod assemblies and torque the connecting rods to 42 ft lb, allowing for 0.0061-inch stretch.

Install the crankshaft bolt in the crankshaft and turn the crankshaft. The rotating assembly should turn over smoothly.

Use a feeler gauge to check the side clearance for each pair of connecting rods.

Lubrication System

Install a stainless-steel 1/4-inch NPT plug inside the vertical drain-back oil line hole in the engine block using pipe thread sealant. Install the oil line fittings for the pressure-out oil line and drain-back oil line using pipe thread sealant.

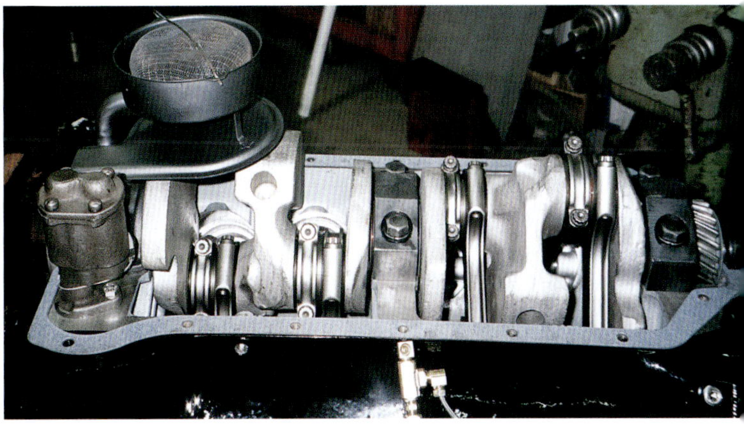

Install a Melling M15 high-volume oil pump with a 1949–1951 Mercury 6615 oil pump cover and inlet tube.

Lubricate the oil pump drive gear with motor oil, and install the oil pump assembly in the engine block. Install the oil pump bolt and lock washer using Loctite, and torque the bolt to 15 ft-lb.

Slide the crankshaft sleeve onto the crankshaft, and gently tap it into position using a piece of steel pipe and a hammer. Slide the one-piece oil seal over the crankshaft sleeve. Install the timing gear cover gasket using silicone sealant, and apply silicone sealant to the inside of the timing gear cover where the oil seal is located. Install the timing gear cover and timing gear cover bolts and lock washers using Loctite on the bolts and torque them to 15 ft-lb.

Install the 1951–1953 Ford C8RT-6059-B/COBA-6059-A polished-aluminum timing gear cover. The Speedway Motors 912-S12853 one-piece oil seal and Ford 6310 oil slinger are visible.

Place some putty on the oil pump pickup screen, and position the oil pan on the engine block without the oil pan gaskets. Push down firmly on the oil pan, and then remove the oil pan. Measure the compressed thickness of the putty. The proper oil-pan-to-oil-pump-pickup-screen clearance is 3/4 to 1 inch. If the clearance is correct, carry on. If the clearance is not correct, adjust the oil pump pickup screen arm until the correct clearance is obtained. Install the cord-material rear oil seal in the rear of the oil pan by gently tapping it into place with a large socket, or use the head of a large ball-peen hammer and tap the head of the hammer with another hammer. Apply silicone

Install the oil pan. The stainless-steel oil drain lines for the Ardun 1203 engine block exhaust port block-off plates are shown on the left side.

sealant around the one-piece oil seal (where it fits in the oil pan), and install the oil pan gaskets using silicone sealant. Install the oil pan and oil pan bolts, lock washers, and AN flat washers using Loctite and torque the bolts to 15 ft-lb. Screw the dipstick tube into the oil pan using silicone sealant, and install the dip stick.

Install the oil line from the pressure-out fitting in the engine block to the inlet side of the oil filter adapter using pipe thread sealant. Install the oil filter line from the drain-back line in the engine block to the outlet side of the oil filter adapter using pipe thread sealant. Fill the oil filter with motor oil and install it on the oil filter adapter.

Pour the required amount of oil in the engine, and install a temporary oil-pressure gauge.

Install the engine block exhaust port gaskets using high-temperature silicone sealant. Install the exhaust port block-off plates and exhaust port block-off plate bolts and lock washers using Loctite, and torque them to 20 ft-lb. Install the exhaust port block-off plate oil drain line fittings, the oil pan drain line fittings, and the exhaust port block-off plate oil drain lines using pipe thread sealant.

Install the cylinder head oil drain line fittings in the cylinder heads and oil pan using pipe thread sealant. Do not install the cylinder head oil drain lines until the exhaust headers have been installed.

Camshaft and Cylinder Heads

After the cylinder heads are "surgically" clean, install Ardun valve spring seat cups over the valve guides and gently tap them into position using a large socket and hammer. Apply motor oil to the valve stem of a valve, and install it

Install the Ardun 1024 milk-bottle-shaped second-design valve lifters. Three of the four set screws used to plug the cylinder head bolt holes at the top of the deck surface are visible.

in the cylinder head. Install the valve stem seal over the valve and around the top of the valve guide. Lubricate the valve spring with motor oil, and install it over the valve and onto the valve spring seat cup. Lubricate the valve spring retainer with motor oil, and install it on the valve and onto the valve spring. Use an OHV valve spring compressor to compress the valve spring and install the valve stem locks. The valve springs must be installed at the manufacturer's recommended height; shim the valve springs as required. Install the remaining valves using the procedure described above.

Lubricate the body of each intake and exhaust valve lifter (tappet) with motor oil and the face of each valve lifter (tappet) with the camshaft assembly lubricant supplied by the camshaft manufacturer. Install all the valve lifters (tappets). Place some putty on the number 1 piston top over the notches for the intake and exhaust valves. Install four cylinder head studs in the right (passenger's) side of the engine block (one at each corner of the engine block) using Permatex aviation form-a-gasket. Install the cylinder head gasket, and slide the assembled cylinder head over the cylinder head studs onto the engine block. Install the intake and exhaust valve pushrods for the number 1 cylinder. Lift the rocker arm assembly over the cylinder head studs onto the cylinder head. Install the four cylinder head stud nuts and flat washers using molykote, and snugly tighten them. Adjust the valve lash for the number 1 cylinder according to the camshaft manufacturer's specifications (timing tag). Rotate the crankshaft until the exhaust valve for the number 1 cylinder starts to open, and adjust the valve lash for the number 1 intake valve. Rotate the crankshaft until the intake valve for the number 1 cylinder starts to close, and adjust the valve lash for the number 1 exhaust valve. Rotate the crankshaft at least two full turns. Remove the rocker arm assembly, the two pushrods, and the cylinder head. Measure the thickness of the compressed putty. The absolute minimum intake-valve-to-piston-top clearance is 0.100 inch, and the absolute minimum exhaust-valve-to-piston-top clearance is 0.125

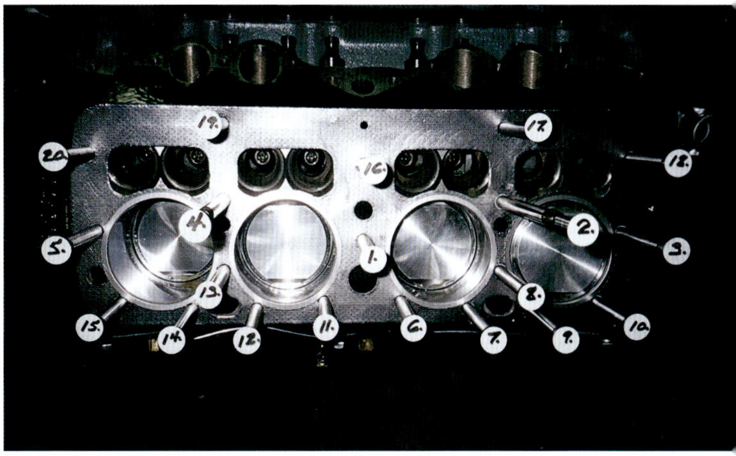

Install the Ardun Enterprises (ARP) cylinder head stud kit. The number 6, 10, 11, 15, 18, and 20 studs are the short studs. The number 2 and 4 studs are the long valve cover/cylinder head studs. The rest of the studs are medium-length studs.

This is Ardun Mechanical Corporation's recommended torque sequence for the Ardun aluminum cylinder heads. Torque the cylinder head stud nuts to 50 ft-lb.

inch (with the cylinder head gasket) when using forged steel connecting rods. The Ardun engine described in this book has an intake-valve-to-piston-top clearance of 0.195 inch and an exhaust-valve-to-piston-top clearance of 0.205 inch (with the cylinder head gasket).

Install the cylinder head studs in the engine block in their correct positions using Permatex aviation form-a-gasket. Slide the cylinder head gaskets over the cylinder head studs and onto the engine block deck. Do not use cylinder head gasket sealant with the Victor/Reinz nitroseal cylinder head gaskets. Some of the cylinder head studs might have to be tweaked slightly with a piece of steel pipe in order to correctly align them when installing the cylinder head gaskets. Apply antiseize compound to the shafts of all the cylinder head studs.

Slide the cylinder heads over the cylinder head studs and onto the cylinder head gaskets. Install the pushrods and apply molykote to the rocker end of each pushrod. Slide the rocker arm assemblies over the cylinder head studs and onto the cylinder head. Apply molykote to the threads on the ends of all the cylinder head studs, and install the cylinder head stud washers and cylinder head stud nuts. Torque the cylinder head stud nuts to 50 ft-lb using the recommended torque sequence.

Adjust the valve lash for all the valves. This procedure is explained in Paragraph 2 above.

Install the rocker arm lubrication fittings and lines using pipe thread sealant. Install the cylinder head lubrication line fittings and lines using pipe thread sealant.

Gap the spark plugs at 0.050 inch, and coat the threads with antiseize compound. Install the spark plugs in the cylinder heads, and torque them to 15 ft-lb. Install the spark plug tubes in the cylinder heads using antiseize compound.

Install the Ardun valve cover stud in each cylinder head using antiseize compound. Install the neoprene rubber

Install the rocker arm assembly and torque the cylinder head stud nuts to 50 ft-lb. Notice the custom rocker arm locating springs.

This picture shows the stainless-steel oil feed lines and T valves for the rocker arm shafts from a pressure-out fitting at the rear of the engine block.

valve cover gaskets on the valve covers using silicone sealant, and then install the valve covers on the cylinder heads dry; do not use any sealant. Secure the valve covers in position with the three acorn nuts and AN flat washers (for each cylinder head) using antiseize compound.

Install the polished stainless-steel spark plug tubes using antiseize compound.

Slip the neoprene rubber O-rings on the serrated spark plug tube nuts. Install the serrated spark plug tube nuts on the spark plug tubes using antiseize compound and tighten them snugly.

Trim the valley cover gasket to the shape of the valley cover. Install the valley cover gasket using silicone sealant. Install the valley cover, valley cover bolts, and AN flat washers using Loctite, and torque them to 25 ft-lb. Install the valley cover chrome acorn nut covers.

Install the rear Ardun blanking cover gaskets on the cylinder heads using silicone sealant. Install the blanking cover bolts and lock washers using antiseize compound, and torque them to 15 ft-lb. Install the neoprene rubber O-rings on the water outlet elbows, and position the water outlet elbows on the front of the cylinder heads. Install the bolts and lock washers using antiseize compound, and torque them to 15 ft-lb.

Intake System

Apply Permatex aviation form-a-gasket to the intake-manifold-to-blower gasket, and slide the gasket over the blower studs onto the blower case. Install the blower assembly on the intake manifold, and turn the complete assembly

Luke's Custom Machine & Design fabricated this simple and effective pushrod/rocker arm lubrication system.

The blower and intake manifold are upside down, in order to install the blower stud neoprene rubber O-rings and the stud nuts.

Luke's Custom Machine & Design fabricated this right-hand polished-aluminum water outlet elbow, located behind and above the distributor.

upside down on a clean towel. Slide the neoprene rubber O-rings onto the blower studs, and install the flat washers and nylocks on the blower studs using pipe thread sealant. Torque the blower studs to 25 ft-lb, and turn the complete assembly right-side up. Slide the intake runner flanges onto the intake runners (the beveled side faces outward), followed by the intake runner flange O-rings. The four intake runner flanges that are notched are for the left (driver's) side of the engine block. Install the intake manifold studs in the left (driver's) side cylinder head only using antiseize compound. Place the intake manifold in position on the cylinder heads, and install the intake manifold studs in the other cylinder head using antiseize compound. Check for adequate clearance between the bottom of the blower studs and the valley cover. The chrome acorn nut covers on the valley cover bolts might not clear the underside of the intake manifold. If that is the case, leave them off. Install the acorn nuts and AN flat washers using antiseize compound, and torque the acorn nuts from the center outward to 25 ft-lb.

Install the crankcase breather/fuel pump stand housing studs in the valley cover using antiseize compound. Install the crankcase breather/fuel pump stand housing gasket using silicone sealant. Bolt the crankcase breather/fuel pump stand housing in position with the acorn nuts and lock washers using Loctite, and torque them to 15 ft-lb. Install the crankcase breather/oil filler cap. Pour enough gear oil into the blower front cover so it is about halfway up the oil-level sight glass.

Install the blower snout assembly gasket using Permatex aviation form-a-gasket. Install the blower snout assembly and blower snout bolts using antiseize compound, and torque them to 20 ft-lb. Install the blower snout drive pulley and blower snout drive pulley bolts, lock washers, and AN flat washers using antiseize compound, and torque them to 25 ft-lb.

Install the blower-to-carburetor-adapter gasket using Permatex aviation form-a-gasket. Place the carburetor adapter on the blower, and install the carburetor adapter bolts using antiseize compound, and torque them to 20 ft-lb. Install the carburetor studs using antiseize compound.

Install the Austin Model BB blower intake manifold and GMC 4-71 blower.

Install the carburetor with the carburetor gasket and then the carburetor stud nuts and AN flat washers using antiseize compound and tighten them snugly; do not overtighten them. Install the dual-inlet fuel line in the carburetor. Install the fuel-pressure gauge in the fuel-pressure regulator, and install this unit in the end of the carburetor fuel line.

Install the blower belt idler pulley on the blower belt idler pulley bracket using antiseize compound on the bolt. Install the blower belt idler pulley bracket on the front of the valley cover using antiseize compound on the bolt.

Ignition System

Position the crankshaft pulley Woodruff key in the crankshaft slot using antiseize compound, and gently tap the Woodruff key into position using a hammer. Apply antiseize compound to the snout of the crankshaft, and install the crankshaft pulley. The pulley should have a snug fit, but it should not require the use of a harmonic balancer installation tool. Gently tap the crankshaft pulley onto the crankshaft snout using a piece of wood and a hammer. Install the crankshaft pulley bolt and washer using Loctite, and torque the crankshaft bolt to 50 ft-lb. Install the

Install a Holley 0-4776C double-pumper carburetor with a Blower Drive Service 14B4 carburetor adapter, along with a Holley 12-803 fuel-pressure regulator and Holley 26-500 fuel-pressure gauge.

Luke's Custom Machine & Design fabricated this stainless-steel distributor hold-down bracket for use with a 1949–1953-style distributor in an Ardun engine.

crankshaft blower drive pulley on the front of the crankshaft pulley using antiseize compound on the bolts, lock washers, and AN flat washers, and torque them to 25 ft-lb.

Rotate the crankshaft until the engine is set with the number 1 piston at 10 degrees before top dead center (TDC), and insert the distributor in the timing gear cover with the distributor rotor at the number 1 spark plug lead position. Clamp the distributor down.

Lay out the spark plug wire, and cut it to length for each cylinder. The cylinders on the right (passenger's) side of the engine are numbered 1, 2, 3, and 4 from the front of the engine block. The cylinders on the left (driver's) side of the engine are numbered 5, 6, 7, and 8 from the front of the engine block. The firing order is 1, 5, 4, 8, 6, 3, 7, 2. Assemble the Hemi spark plug tube and insulator kit on the end of the spark plug wire for each cylinder (this installation is explained in Chapter 27), and connect them to the spark plugs. Slide the Hemi spark plug tube covers onto the serrated spark plug tube nuts. Install the distributor cap nipples and terminals on the ends of the spark plug wires, and insert them in the correct firing order in the distributor cap. Install the ignition coil and ignition coil bracket using antiseize compound on the bolt. Connect the ignition coil wire to the distributor.

Install the 1949–1953 Ford 6375 flywheel, Centerforce 280490 clutch disc, and Ford 7563 long-style pressure plate.

Install the generator mounting brackets using antiseize compound on the bolts. Install the generator on the generator mounting brackets.

Install the crankshaft/water pumps/generator idler bracket and pulley using antiseize compound on the bolts.

The flywheel or flexplate will have to be installed in order to start the engine. Install the flywheel or flexplate bolts using Loctite, and torque them to 80 ft-lb. A bell housing and starting motor plate will have to be installed in order to mount the starting motor. Install the starting motor bolts using antiseize compound.

Bolt the Powermaster 9507 XS torque starting motor in place.

Cooling System and Miscellaneous

Install the water pump gaskets using silicone sealant. Install the water pumps and water pump bolts and lock washers using Loctite, and torque them to 25 ft-lb.

Install the crankshaft/water pumps/generator V-belt, allowing for approximately 3/4 to 1 inch slack.

Install the blower drive belt, allowing for approximately 3/4 inch slack.

Bolt on the headers with mufflers, and connect the drain lines from the cylinder heads to the oil pan. Install a radiator with some water in it, wire up the ignition system, and hook up a battery with a starter switch. Start the engine, and adjust the timing and the carburetor. Allow the engine to run at approximately 2,000 rpm for 30 minutes in order to break in the camshaft. After the engine is

This picture shows the rear of the Ardun engine with the bell housing bolted in place. The sorry-looking exhaust headers are only used for the initial startup of the engine.

at full operating temperature, shut it off and allow the engine to cool. Remove the valve covers, and retorque the cylinder heads using the approved torque pattern. Install the valve covers, and retorque the valve cover acorn nuts.

This engine has a smooth sound to it and a nice torque curve from 2,000 to 4,000 rpm. This is a sure indication that the correct camshaft was selected for this particular engine.

The front of the Ardun engine looks fantastic with all the pulleys and belts installed. I am certain even Zora Arkus-Duntov would have admired this engine.

This is the right side of the completed Ardun engine. It is truly a magnificent piece of engineering, and it runs as well as it looks.

APPENDIX A
DYNO PRINTOUTS

263-ci Ford flathead V-8 Model 59 street-performance engine

Note: The estimated output of this engine is 235 horsepower at 5,500 rpm and 269 ft-lb torque at 3,500 rpm.

This is the 263-ci Ford flathead V-8 engine described in Section I of this book. Luke's Custom Machine & Design built the engine for the owner, who installed it in a daily driven 1940 Ford Coupe.

SHORT BLOCK

Block:	FORD "FLATHEAD".	Bore:	3.343 in	Stroke:	3.750 in
Cylinders:	8	Cyl Vol:	539.4 cc	Total Vol:	263.3 ci

CYLINDER HEADS

Cylinder Heads:	Low Performance/Stock Ports And Valves		
Air Flow File:	***		
Intake Valves:	1	Exhaust Valves:	1
Intake Valve:	1.610 in	Exhaust Valve:	1.610 in

COMPRESSION

Compression Ratio:	9.10	Combustion Space:	66.6 cc

INDUCTION

Induction Flow:	486.0 cfm @ 1.5 inHg	Fuel:	Gasoline
Manifold Type:	Dual-Plane Manifold	N20:	0.0 lbs/min

Blower:	None			Intercooler:	*** %
Flow:	*** cfm	Pressure Ratio:	***	Boost Limit:	*** psi
Speed:	*** rpm	Belt Gear Ratio:	***	Surge Flow:	*** cfm
Eff:	*** %	Internal Gear Ratio:	***		

EXHAUST

Exhaust System:	H.P. Manifolds And Mufflers

CAMSHAFT

Camshaft Type:	ISKY "MAX #1".	Cam File:	***
Lifter:	Solid	Lobe Center:	110.0
Cam Specs @:	0.050-Lift	Valve Overlap:	6.0
Int Lift@Valve:	0.350 in	Int Duration:	226.0
Exh Lift@Valve:	0.350 in	Exh Duration:	226.0
Nominal Timing		Timing@ Adv(+)/Ret(-):	0.0

IVO (BTDC):	3.0	IVC (ABDC):	43.0	IVO:	3.0	IVC:	43.0
EVO (BBDC):	43.0	EVC (ATDC):	3.0	EVO:	43.0	EVC:	3.0
ICA (ATDC):	110.0	ECA (BTDC):	110.0	ICA:	110.0	ECA:	110.0

CALCULATED POWER AND ENGINE PRESSURES

Engine RPM	Power (Fly)	Torque (Fly)	Int Man Pressure	Vol Eff %	IMEP Pressure	FMEP Pressure	BMEP Pressure
2000	94	246	14.66	67.6	186.4	17.8	142.9
2500	120	252	14.65	69.8	183.0	19.3	146.2
3000	151	265	14.62	74.1	189.4	21.0	153.8
3500	179	269	14.57	77.3	189.9	22.8	156.2
4000	203	266	14.52	79.2	188.5	24.7	154.5
4500	222	259	14.46	79.9	186.2	26.7	150.4
5000	232	244	14.40	79.4	179.1	28.9	141.7
5500	235	224	14.35	78.0	169.3	31.2	130.3
6000	223	195	14.29	74.3	153.8	33.6	113.4
6500	203	164	14.27	70.2	137.3	36.2	95.4
7000	175	131	14.25	65.7	119.8	38.8	76.4
7500	147	103	14.25	61.5	105.2	41.6	59.9
8000	108	71	14.25	57.1	88.4	44.6	41.3
8500	72	44	14.26	53.2	75.0	47.6	25.8
9000	28	17	14.27	49.4	61.0	50.8	9.6
9500	0	0	14.29	45.7	47.8	54.1	-6.0
10000	0	0	14.31	42.4	36.6	57.6	-19.8
10500	0	0	14.33	39.1	24.9	61.2	-34.3
11000	0	0	14.36	36.3	15.8	64.9	-46.3

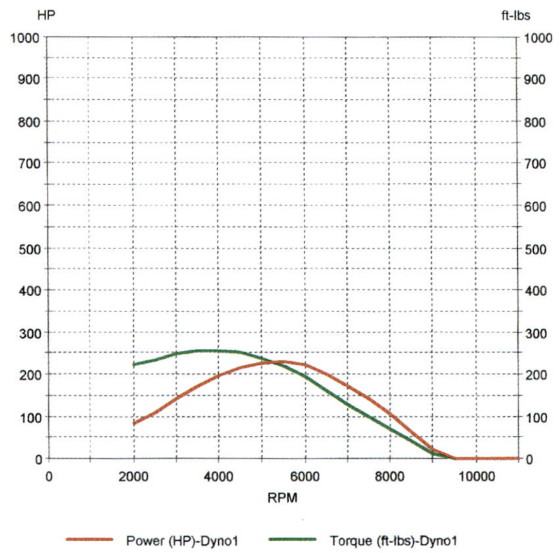

Blown 262-ci French military flathead V-8 street-performance engine

Note: The estimated output of this engine is 281 horsepower at 5,000 rpm and 323 ft-lb torque at 3,500 rpm.

SHORT BLOCK

Block: FRENCH "FLATHEAD" | Bore: 3.228 in | Stroke: 4.000 in
Cylinders: 8 | Cyl Vol: 536.4 cc | Total Vol: 261.9 ci

CYLINDER HEADS

Cylinder Heads: Low Performance/Stock Ports And Valves
Air Flow File: ***
Intake Valves: 1 | Exhaust Valves: 1
Intake Valve: 1.720 in | Exhaust Valve: 1.500 in

COMPRESSION

Compression Ratio: 8.00 | Combustion Space: 76.6 cc

INDUCTION

Induction Flow: 600.0 cfm @ 1.5 inHg | Fuel: Gasoline
Manifold Type: Forced Induction | N20: 0.0 lbs/min

Blower: Roots- BDS 471 (Street) | Intercooler: *** %
Flow: 260.0 cfm | Pressure Ratio: *** | Boost Limit: 8.2 psi
Speed: *** rpm | Belt Gear Ratio: 0.86 | Surge Flow: *** cfm
Eff: 55.0 % | Internal Gear Ratio: ***

EXHAUST

Exhaust System: Forced Induction Exhaust

CAMSHAFT

Camshaft Type: CROWER BLOWER. | Cam File: ***
Lifter: Solid | Lobe Center: 112.0
Cam Specs @: 0.050-Lift | Valve Overlap: 0.0
Int Lift@Valve: 0.321 in | Int Duration: 224.0
Exh Lift@Valve: 0.319 in | Exh Duration: 224.0
Nominal Timing | Timing@ Adv(+)/Ret(-): 0.0

IVO (BTDC): 4.0 | IVC (ABDC): 40.0 | IVO: 4.0 | IVC: 40.0
EVO (BBDC): 48.0 | EVC (ATDC): -4.0 | EVO: 48.0 | EVC: -4.0
ICA (ATDC): 108.0 | ECA (BTDC): 116.0 | ICA: 108.0 | ECA: 116.0

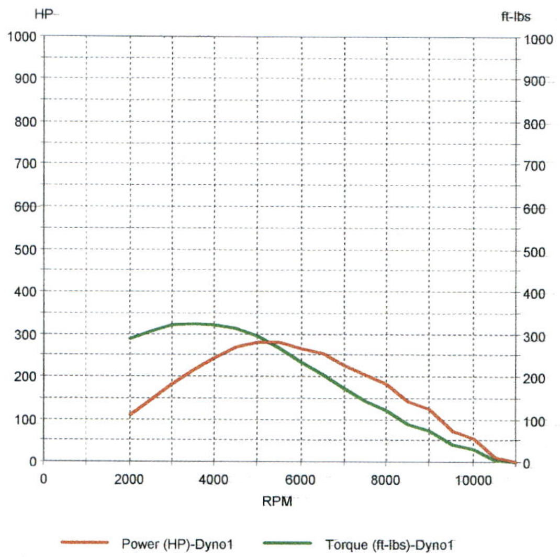

This is the blown 262-ci French military flathead V-8 engine described in Section II of this book. Luke's Custom Machine & Design built this engine for the owner, who installed it in a daily driven 1934 Ford pickup truck.

Engine RPM	Power (Fly)	Torque (Fly)	Int Man Pressure	Vol Eff %	IMEP Pressure	FMEP Pressure	BMEP Pressure
2000	110	289	14.96	78.3	217.0	18.2	168.5
2500	146	307	15.48	84.1	220.5	19.9	179.1
3000	182	319	16.00	89.2	225.6	21.7	186.2
3500	215	323	16.51	93.3	225.5	23.6	188.7
4000	243	319	17.01	95.7	223.4	25.7	186.5
4500	268	313	17.51	97.2	221.6	28.0	182.6
5000	281	295	18.00	96.7	212.9	30.4	172.2
5500	279	266	18.49	94.0	197.5	32.9	155.3
6000	266	233	19.00	89.8	179.8	35.6	136.0
6500	253	205	19.52	86.1	165.2	38.5	119.6
7000	226	170	20.04	81.1	146.5	41.4	99.1
7500	202	142	20.58	76.5	132.3	44.6	82.7
8000	182	119	21.13	73.3	121.6	47.8	69.6
8500	143	88	21.66	68.3	105.8	51.3	51.4
9000	124	72	22.22	65.9	99.6	54.8	42.3
9500	78	43	22.47	61.0	85.1	58.5	25.0
10000	51	27	22.49	57.8	79.1	62.4	15.8
10500	12	6	22.50	54.3	70.2	66.4	3.6
11000	0	0	22.51	50.7	61.4	70.5	-8.6

Blown 284-ci Ardun V-8 street-performance engine

Note: The estimated output of this engine is 377 horsepower at 6,000 rpm and 375 ft-lb torque at 4,000 rpm.

This is the blown 284-ci Ardun V-8 engine described in Section III of this book. Luke's Custom Machine & Design built the engine for the owner, who installed it in a daily driven 1940 Mercury coupe.

SHORT BLOCK

Block:	BLOWN ARDUN.	Bore:	3.313 in	Stroke:	4.125 in
Cylinders:	8	Cyl Vol:	582.7 cc	Total Vol:	284.5 ci

CYLINDER HEADS

Cylinder Heads:	Canted/Rectangular Ports, Fully Ported		
Air Flow File:	***		
Intake Valves:	1	Exhaust Valves:	1
Intake Valve:	1.750 in	Exhaust Valve:	1.500 in

COMPRESSION

Compression Ratio:	8.00	Combustion Space:	83.2 cc

INDUCTION

Induction Flow:	600.0 cfm @ 1.5 inHg	Fuel:	Gasoline
Manifold Type:	Forced Induction	N20:	0.0 lbs/min

Blower:	Roots- BDS 471 (Street)			Intercooler:	*** %
Flow:	260.0 cfm	Pressure Ratio:	***	Boost Limit:	7.3 psi
Speed:	*** rpm	Belt Gear Ratio:	0.94	Surge Flow:	*** cfm
Eff:	55.0 %	Internal Gear Ratio:	***		

EXHAUST

Exhaust System:	Forced Induction Exhaust

CAMSHAFT

Camshaft Type:	CROWER BLOWER.	Cam File:	***
Lifter:	Solid	Lobe Center:	112.0
Cam Specs @:	0.050-Lift	Valve Overlap:	-2.0
Int Lift@Valve:	0.386 in	Int Duration:	222.0
Exh Lift@Valve:	0.384 in	Exh Duration:	222.0
Nominal Timing		Timing@ Adv(+)/Ret(-):	0.0

IVO (BTDC):	3.0	IVC (ABDC):	39.0	IVO:	3.0	IVC:	39.0
EVO (BBDC):	47.0	EVC (ATDC):	-5.0	EVO:	47.0	EVC:	-5.0
ICA (ATDC):	108.0	ECA (BTDC):	116.0	ICA:	108.0	ECA:	116.0

Engine RPM	Power (Fly)	Torque (Fly)	Int Man Pressure	Vol Eff %	IMEP Pressure	FMEP Pressure	BMEP Pressure
2000	127	334	15.16	81.7	230.5	18.4	179.8
2500	169	355	15.73	87.6	233.7	20.1	190.7
3000	211	370	16.29	93.0	239.8	22.0	198.8
3500	249	374	16.84	96.6	239.1	24.1	200.9
4000	286	375	17.38	99.4	239.9	26.3	201.5
4500	318	371	17.91	100.5	240.0	28.7	199.4
5000	345	363	18.43	101.0	237.8	31.2	194.9
5500	362	345	18.95	100.2	230.7	33.8	185.7
6000	377	330	19.48	99.5	224.5	36.7	177.2
6500	375	303	20.00	97.0	212.3	39.7	162.9
7000	367	276	20.54	94.2	199.8	42.8	148.1
7500	355	248	21.09	90.9	187.6	46.1	133.5
8000	331	217	21.20	87.0	173.4	49.5	116.9
8500	296	183	21.17	81.6	157.4	53.1	98.4
9000	268	156	21.18	77.4	145.9	56.9	84.0
9500	229	127	21.18	72.8	133.0	60.8	68.1
10000	196	103	21.19	68.9	123.5	64.9	55.3
10500	158	79	21.20	65.7	114.2	69.1	42.5
11000	106	51	21.20	61.5	102.4	73.5	27.3

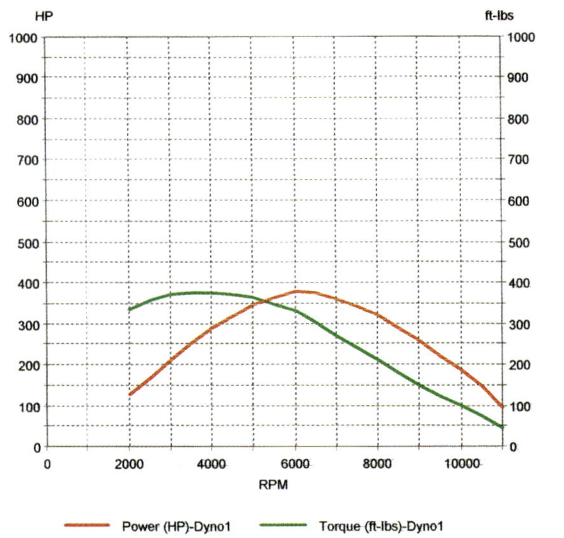

APPENDIX B
RESOURCES

Listed below, in alphabetical order, are the parties and parts sources I believe are important to know when building a Ford flathead V-8 engine. The companies with brief explanations are the companies I presently deal with on a regular basis. (Note: I have not received any commercial or financial considerations from those companies in order to mention them. This is most unfortunate, because I do not think an all-expense-paid trip from any of those companies for me to attend the annual Munich beer festival [on a regular basis] would be out of order!)

There are a number of excellent speed equipment manufacturers and engine building shops across North America. Some of these facilities have been mentioned in other publications and hot rod magazine articles pertaining to the Ford flathead V-8 engine. I have not mentioned them in this book because I have not used their services in building any of my own engines, and I am hesitant to recommend a company with which I have had no experience. I do not intend it as a snub of anyone or any business.

There are hundreds of manufacturers and parts outlets positioned around North America, and it would be impossible to include them all in this book. I have mentioned a few of those outlets. Most readers will have their own sources for parts.

Ardun Enterprises
Mr. Don Ferguson
P.O. Box 205
Wilmington, California 90744
Phone: (323) 775-6803
Fax: (310) 834-9904
Website: www.ardun.com

The Rod Shop
Mr. Ken Austin
P.O. Box 111
2601 Crestview Drive
Newburg, Oregon 97132
Phone: (503) 349-1360
Fax: (503) 537-2760

Blower Drive Service
12140 E. Washington Boulevard
Whittier, California 90606
Phone: (562) 693-4302
Fax: (562) 696-7091
Website: www.blowerdriveservice.com

Crane Cams, Inc.
530 Fentress Boulevard
Daytona Beach, Florida 32114
Phone: (386) 258-1151
Tech Line: (386) 258-6174
Fax: (386) 258-6167
Website: www.cranecams.com

Crower Cams & Equipment Company, Inc.

I have been buying camshafts and valvetrain components only from Crower Cams & Equipment Company for several years. I believe that company produces the finest-quality camshafts and components at an affordable price.

Jerry MacLaughlin, of the technical department, has been extremely helpful and courteous with questions I have had over the years relating to camshafts for specific applications. Jerry recommended the camshafts used in the engines described in Sections II and III of this book. These camshafts really perform well.

Crower Cams & Equipment Company, Inc.
3333 Main Street
Chula Vista, California 91911-5899
Phone: (619) 661-6477
Fax: (619) 661-6466
Website: www.crower.com

Edelbrock Corporation
2700 California Street
Torrence, California 90503
Phone: (310) 781-2222
Fax: (310) 320-1187
Website: www.edelbrock.com

Egge Parts House
11707 Slausen Avenue
Sante Fe Springs, California 90670
Toll-free: (800) 886-3443
Phone: (562) 945-3419
Fax: (562) 693-1635
Website: www.egge.com
E-mail: info@egge.com

Hedman Street Hedders

16410 Manning Way
Cerritos, California 90703
Phone: (562) 921-0404
Fax: (562) 921-7515
Website: www.hedman.com
E-mail: techsupport@hedman.com

High Performance Engines Ltd.

Bud Child founded High Performance Engines about 40 years ago, and Dave Child, his son, now owns and operates the company. Although supposedly retired, Bud is usually at the shop six days a week, working 10-hour days at sweatshop wages for Dave.

I have been using the services of that company for the machine work on my own engines for over 30 years. High Performance Engines builds engines for customers all over the Pacific Northwest, and many of those engines have produced drag racing and drag boat winners.

My thanks to Dave Child for allowing me access to his shop in order to take the photographs of the machining procedures shown in this book.

High Performance Engines Ltd.

4329 Buchanan Street
Burnaby, B.C. V5C 3X7
Canada
Phone: (604) 299-6131
Fax: (604) 299-6017

Iskenderian Racing Cams

18020 South Broadway, P.O. Box 30
Gardena, California 90247-9990
Phone: (323) 770-0930
Fax: (310) 515-5730
Website: www.iskycams.com

Jitney Auto Parts Ltd.

Keith Lee is the owner and operator of Jitney Auto Parts. He is one of the most knowledgeable flathead persons I know of and a true gentleman. Jitney Auto Parts will ship orders to just about anywhere in the world. Keith stocks all the brand-name Ford flathead V-8 speed equipment and has a large stock of used vintage high-performance and stock flathead parts, including transmissions and rear ends.

Jitney Auto Parts Ltd.

Mr. Keith Lee
P.O. Box 23048, Sub 31
Saskatoon, Saskatchewan S7J 5H3
Canada
Phone: (306) 668-3673
Fax: (306) 668-3658

Doug King Enterprises

Mr. Doug King
Phone: (510) 537-3909

Luke's Custom Machine & Design Ltd.

Luke Balogh is the owner and operator of Luke's Custom Machine & Design. He has operated that business for over 25 years, although he has been building high-performance engines and custom automobiles for over 30 years. Luke is an accomplished drag racer who has raced at the Bonneville salt flats, setting an XX/Street Roadster record there in 1986 with an Ardun-powered 1927 Ford Model T roadster.

Luke is renowned for his aluminum designs, fabrication, and welding. He produces complete blower setups for just about any engine. Luke has built a variety of engines over the years, some of which have been featured in hot rod magazines and other publications. His preference is for the Ford flathead V-8, the Ardun, and the early Chrysler Hemi engines.

Luke's Custom Machine & Design Ltd.

Mr. Luke Balogh
1457 Charlotte Road
North Vancouver, B.C. V7J 1H1
Canada
Phone: (604) 980-8617
Fax: (604) 980-8656

Manley Performance Parts

1960 Swarthmore Avenue
Lakewood, New Jersey 08701
Phone: (732) 905-3366
Fax: (732) 905-3010
Website: www.manleyperformance.com

Mooneyes USA, Inc.

10820 South Norwalk Boulevard
Sante Fe Springs, California 90670
Phone: (562) 944-6311
Fax: (562) 946-2961
Website: www.mooneyes.com

Offenhauser Sales Corporation

P.O. Box 32218
Los Angeles, California 90032
Phone: (323) 225-1307

Pacific Fasteners, Inc.
Pacific Fasteners is one of the best sources for any type of high-quality stainless-steel fastener. This company appears to stock everything that is required for an engine, or even an entire automobile. You can buy fasteners individually or in bulk quantities; however, there is usually a $5 minimum for orders. Pacific Fasteners will ship FedEx or UPS to just about anywhere in the world.

Pacific Fasteners, Inc.
2411 South 200 Street
Seattle, Washington 98198
Phone: (206) 824-0416
Fax: (206) 878-1041
Website: www.pacificfasteners.com
E-mail: pacfast@pacificfasteners.com

Pacific Fasteners Ltd.
3934 East 1st Avenue
Burnaby, B.C. V5C 5S3
Canada
Phone: (604) 294-9411
Fax: (604) 294-4730

Redi-Strip Metal Cleaning Canada Ltd.
Redi-Strip Metal Cleaning Canada has been in business for over 20 years. This company specializes in metal cleaning, from engine blocks to entire automobile body shells. They do excellent work and their prices are competitive.

Redi-Strip Metal Cleaning Canada
7961 Vantage Way
Delta, B.C. V4G 1A6
Canada
Phone: (604) 946-7761
Fax: (604) 946-5936

Red's Headers
Mr. Red Hamilton
22950 Bednar Lane
Fort Bragg, California 95437
Phone: (707) 964-7733
Website: www.redsheaders.com
E-mail: info@redsheaders.com

Ross Racing Pistons
Ross Racing Pistons produces top-quality forged aluminum pistons using state-of-the-art CNC machining technology, and the company offers some of the best pistons money can buy at an affordable price. Ross Racing Pistons manufactures pistons in various sizes, from lawn mowers to aircraft, and they can custom build just about any type of piston.

Ross Racing Pistons
625 South Douglas
El Segundo, California 90245
Toll-free: (800) 392-7677
Tech Line: (310) 536-0100
Fax: (310) 536-0333
Website: www.rosspistons.com

RPM Catalog (Dale Wilch's)
Dale Wilch is the owner of RPM Catalog, and he has been selling used speed equipment for years. I have purchased items from him in the past, and I have found his service to be prompt and efficient. The small-block Chevrolet stainless-steel 1.61-inch-diameter valves are ideal for the Ford flathead V-8 engine and are available at a bargain price.

RPM Catalog
Mr. Dale Wilch
P.O. Box 12301
Kansas City, Kansas 66112
Phone: (913) 788-3219
Fax: (913) 788-9682
Website: www.rpmcat.com

Sanderson Headers
517 Railroad Avenue
South San Francisco, California 94080
Phone: (650) 583-6617
Fax: (650) 583-8475
Website: www.sandersonheaders.com
E-mail: sales@sandersonheaders.com

SCAT Enterprises, Inc.
1400 Kingsdale Avenue
Redondo Beach, California 90278
Phone: (310) 370-5501
Fax: (310) 214-2285
Website: www.scatenterprises.com

Schneider Racing Cams, Inc.
1235 Cushman Avenue
San Diego, California 92110
Phone: (619) 297-0227
Fax: (619) 297-0577
Website: www.schneidercams.com
E-mail: schneiderracingcams@sbcglobal.net

Smith Brothers Pushrods
1320 S.E. Armour Road, A-1
Bend, Oregon 97702
Toll-free: (800) 367-1533
Phone: (541) 388-8188
Fax: (541) 389-8840
Website: www.pushrods.net
E-mail: smithbros@bendcable.com

Speedway Motors, Inc.
I have been buying Ford flathead V-8 engine and hot rod parts for my own use from Speedway Motors since 1983, and they have always been polite and efficient when filling my orders. This is one of the best sources for flathead speed equipment.

Speedway Motors, Inc.
P.O. Box 81906
Lincoln, Nebraska 68501
Phone: (402) 323-3200
Fax: (402) 323-3211
Website: www.speedwaymotors.com

Summit Racing Equipment
P.O. Box 909
Akron, Ohio 44309-0909
Toll-free: (800) 230-3030
Fax: (330) 630-5333
Website: www.summitracing.com

Tatom Custom Engines
P.O. Box 2504
Mount Vernon, Washington 98273
Phone: (360) 424-8314
Fax: (360) 424-6717
Website: www.tatom.com
E-mail: flatheads@tatom.com

Wilcap Company
P.O. Box 763
Pismo Beach, California 93448
Phone: (805) 481-7639
Website: www.wilcap.com

Wilkinson's Automobilia
Wilkinson's Automobilia is one of the finest sources of vintage and current automotive books, shop manuals, chassis manuals, and hot rod magazines. Some of the publications go back to the early part of the twentieth century, and they seem to have everything. These people are extremely helpful and very friendly.

Wilkinson's Automobilia
2531 Ontario Street
Vancouver, B.C. V5T 2X7
Canada
Phone: (604) 873-6242
Fax: (604) 873-6259
Website: *www.eautomobilia.com*
E-mail: info@wilkinsonsauto.com

APPENDIX C
VINTAGE CAMSHAFT SPECIFICATIONS

During the 1940s and 1950s, a number of camshaft grinding and speed-equipment companies offered products for Ford flathead V-8 engines. Almquist Engineering & Manufacturing Company, Cannon Engineering Company, Detroit Racing Equipment Company, Dunn Engineering, Edelbrock Equipment Company, Engle's Machine Shop, Harman & Collins, Howard's Automotive Service, Ed Iskenderian, Potvin Auto Equipment Company, Smith & Jones, Thomas P. Spaulding, Weber Cam Grinding Company, Weiand Racing Equipment, and Winfield provided most of the popular camshafts. Each camshaft company had its own description for the various grinds. The best name I have come across is the "Honest Even Idler" offered by Honest Charley Speed Shop of Chattanooga, Tennessee.

I have accumulated a fair bit of information over the years pertaining to some of these camshafts, which should be shared before it is lost forever. I have not included information for current off-the-shelf camshafts, such as those produced by Iskenderian Racing Cams and Schneider Racing Cams, as that information is readily available. All of the camshafts listed here are sold flat tappet, they were usually regrinds, and most are designed for pre-1949 flathead engines. The duration at 0.050-inch lift is not included with any of these camshafts, as that standard of measurement was not in use during the period of manufacture. The timing events are listed as "advertised duration" because no information was provided as to what amount of lift the timing measurements were taken at. Some of these camshafts lacked the necessary information to produce an accurate picture for comparison purposes. Fortunately, there was enough information to enter into my computer dynamometer simulation program, which resulted in the following camshaft list.

In order to compare these camshafts, I have given the estimated output of the 1948 Ford flathead V-8 engine, described in Section I, when using each of these camshafts. Some of the camshafts are designed for competition use, which requires a larger displacement engine than the one described in Section I, and some are designed to operate continuously at high rpm. I have designated the camshafts as "street" and/or "competition" based on the manufacturer's description and the computer dynamometer simulation results. Many of these vintage camshafts have a large overlap, which will result in higher engine emissions. This could be a problem if you happen to reside in a jurisdiction with mandatory vehicle emissions testing.

A note of caution: I have done my absolute best to provide accurate information for the camshafts listed; however, a lot of this information is over 50 years old and is provided without any guarantee. I have encountered camshafts in the past that had different timing events, even though they were from the same manufacturer and had the same grind number.

Manufacturer: Harman & Collins, Inc.

Harman & Collins, Inc. was an established camshaft grinding company in the 1950s, and it was also noted for manufacturing excellent magnetos. Kenny Harman and Cliff Collins founded the company in Los Angeles, California.

Camshaft number: n/a; Grind: "semi"

Application: street
Advertised duration: 250 degrees intake and exhaust
Lobe separation angle: 109 degrees
Net valve lift: 0.309-inch intake and 0.307-inch exhaust
Valve lash (hot): 0.011-inch intake and 0.013-inch exhaust
Overlap (with advertised duration): 32 degrees
IVO: 16 degrees
IVC: 54 degrees
EVO: 54 degrees
EVC: 16 degrees
Estimated output: 214 horsepower at 4,500 rpm and 285 ft-lb torque at 3,000 rpm

Camshaft number: 6217, 3/4 Race; Grind: "3/4"

Application: street
Advertised duration: 258 degrees intake and 248 degrees exhaust
Lobe separation angle: 111.5 degrees
Net valve lift: 0.309-inch intake and 0.307-inch exhaust

Valve lash (hot): 0.011-inch intake and 0.013-inch exhaust
Overlap (with advertised duration): 30 degrees
IVO: 17 degrees
IVC: 61 degrees
EVO: 55 degrees
EVC: 13 degrees
Estimated output: 211 horsepower at 4,500 rpm and 274 ft-lb torque at 3,000 rpm

Camshaft number: 6014, Full Race; Grind: "FR"

Application: street
Advertised duration: 258 degrees intake and exhaust
Lobe separation angle: 111.5 degrees
Net valve lift: 0.309-inch intake and 0.307-inch exhaust
Valve lash (hot): 0.011-inch intake and 0.013-inch exhaust
Overlap (with advertised duration): 35 degrees
IVO: 17 degrees
IVC: 61 degrees
EVO: 60 degrees
EVC: 18 degrees
Estimated output: 221 horsepower at 5,000 rpm and 273 ft-lb torque at 3,500 rpm

Camshaft number: 6356, Sportsman; Grind: "Sportsman"

Application: street
Advertised duration: 264 degrees intake and exhaust
Lobe separation angle: 113.5 degrees
Net valve lift: 0.376-inch intake and 0.375-inch exhaust
Valve lash (hot): 0.012-inch intake and 0.013-inch exhaust
Overlap (with advertised duration): 37 degrees
IVO: 18 degrees
IVC: 66 degrees
EVO: 65 degrees
EVC: 19 degrees
Estimated output: 234 horsepower at 5,000 rpm and 270 ft-lb torque at 3,500 rpm

Camshaft number: 6015, Super; Grind: "SU"

Application: street and competition
Advertised duration: 270 degrees intake and 258 degrees exhaust
Lobe separation angle: 112 degrees
Net valve lift: 0.313-inch intake and 0.309-inch exhaust
Valve lash (hot): 0.011-inch intake and 0.013-inch exhaust
Overlap (with advertised duration): 40 degrees
IVO: 22 degrees
IVC: 68 degrees
EVO: 60 degrees
EVC: 18 degrees
Estimated output: 220 horsepower at 5,000 rpm and 260 ft-lb torque at 3,500 rpm

Camshaft number: Super "H"; Grind: "SU-H"

Application: competition
Advertised duration: 288 degrees intake and 270 degrees exhaust
Lobe separation angle: 111.5 degrees
Net valve lift: 0.332-inch intake and 0.335-inch exhaust
Valve lash (hot): 0.018-inch intake and 0.015-inch exhaust
Overlap (with advertised duration): 56 degrees
IVO: 30 degrees
IVC: 78 degrees
EVO: 64 degrees
EVC: 26 degrees
Estimated output: 220 horsepower at 5,500 rpm and 233 ft-lb torque at 4,000 rpm

Camshaft number: 6017, Super "M"; Grind: "SU-M"

Application: competition
Advertised duration: 278 degrees intake and 268 degrees exhaust
Lobe separation angle: 109.5 degrees
Net valve lift: 0.339-inch intake and 0.337-inch exhaust
Valve lash (hot): 0.011-inch intake and 0.013-inch exhaust
Overlap (with advertised duration): 58 degrees
IVO: 30 degrees
IVC: 68 degrees
EVO: 68 degrees
EVC: 28 degrees
Estimated output: 228 horsepower at 5,500 rpm and 249 ft-lb torque at 4,000 rpm

Camshaft number: 6016, Super "T"; Grind: "SU-T"

Application: street and competition
Advertised duration: 260 degrees intake and exhaust
Lobe separation angle: 110 degrees

Net valve lift: 0.342-inch intake and 0.340-inch exhaust
Valve lash (hot): 0.010-inch intake and 0.012-inch exhaust
Overlap (with advertised duration): 40 degrees
IVO: 20 degrees
IVC: 60 degrees
EVO: 60 degrees
EVC: 20 degrees
Estimated output: 229 horsepower at 5,000 rpm and 275 ft-lb torque at 3,500 rpm

Manufacturer: Potvin Auto Equipment Company

Chuck Potvin founded Potvin Auto Equipment Company in Anaheim, California. In 1962, Potvin sold the company to Moon Equipment Company, which has continued to grind Potvin camshafts. Moon Equipment Company has been renamed Mooneyes USA (Santa Fe Springs, California).

Camshaft number: 3/4 Race; Grind: "3/4"

Application: street
Advertised duration: 256 degrees intake and exhaust
Lobe separation angle: 111.5 degrees
Net valve lift: 0.316-inch intake and 0.315-inch exhaust
Valve lash (hot): 0.013-inch intake and 0.014-inch exhaust
Overlap (with advertised duration): 33 degrees
IVO: 16 degrees
IVC: 60 degrees
EVO: 59 degrees
EVC: 17 degrees
Estimated output: 222 horsepower at 5,000 rpm a nd 276 ft-lb torque at 3,500 rpm

Camshaft number: Super Race; Grind: "SU"

Application: street
Advertised duration: 270 degrees intake and exhaust
Lobe separation angle: 111.5 degrees
Net valve lift: 0.344-inch intake and 0.343-inch exhaust
Valve lash (hot): 0.013-inch intake and 0.014-inch exhaust
Overlap (with advertised duration): 47 degrees
IVO: 23 degrees
IVC: 67 degrees
EVO: 66 degrees
EVC: 24 degrees
Estimated output: 233 horsepower at 5,500 rpm and 261 ft-lb torque at 3,500 rpm

Camshaft number: Super; Grind: "3/8"

Application: competition
Advertised duration: 280 degrees intake and exhaust
Lobe separation angle: 111.5 degrees
Net valve lift: 0.361-inch intake and 0.360-inch exhaust
Valve lash (hot): 0.014-inch intake and 0.015-inch exhaust
Overlap (with advertised duration): 57 degrees
IVO: 28 degrees
IVC: 72 degrees
EVO: 71 degrees
EVC: 29 degrees
Estimated output: 231 horsepower at 5,500 rpm and 246 ft-lb torque at 4,000 rpm

Camshaft number: Climax; Grind: "400"

Application: competition
Advertised duration: 284 degrees intake and exhaust
Lobe separation angle: 113 degrees
Net valve lift: 0.387-inch intake and 0.386-inch exhaust
Valve lash (hot): 0.013-inch intake and 0.014-inch exhaust
Overlap (with advertised duration): 58 degrees
IVO: 29 degrees
IVC: 75 degrees
EVO: 75 degrees
EVC: 29 degrees
Estimated output: 232 horsepower at 5,500 rpm and 241 ft-lb torque at 4,000 rpm

Camshaft number: Eliminator; Grind: "425"

Application: competition
Advertised duration: 280 degrees intake and exhaust
Lobe separation angle: 114 degrees
Net valve lift: 0.413-inch intake and 0.412-inch exhaust
Valve lash (hot): 0.012-inch intake and 0.013-inch exhaust
Overlap (with advertised duration): 52 degrees
IVO: 26 degrees
IVC: 74 degrees
EVO: 74 degrees
EVC: 26 degrees
Estimated output: 238 horsepower at 5,500 rpm and 249 ft-lb torque at 4,000 rpm
Note: According to Moon Equipment Company, the Potvin "425 Eliminator" camshaft was produced by regrinding a customer-supplied 1932 Ford billet-steel core.

Manufacturer: Smith & Jones

Clay Smith and Dan Jones founded Smith & Jones camshaft-grinding company in Long Beach, California, in the 1940s.

Camshaft number: 85; Grind: "252"

Application: street
Advertised duration: 252 degrees intake and exhaust
Lobe separation angle: 110 degrees
Net valve lift: 0.300-inch intake and 0.299-inch exhaust
Valve lash (hot): 0.012-inch intake and 0.013-inch exhaust
Overlap (with advertised duration): 32 degrees
IVO: 16 degrees
IVC: 56 degrees
EVO: 56 degrees
EVC: 16 degrees
Estimated output: 215 horsepower at 5,000 rpm and 282 ft-lb torque at 3,000 rpm

Camshaft number: 85; Grind: "260"

Application: street
Advertised duration: 260 degrees intake and exhaust
Lobe separation angle: 110 degrees
Net valve lift: 0.300-inch intake and 0.299-inch exhaust
Valve lash (hot): 0.012-inch intake and 0.013-inch exhaust
Overlap (with advertised duration): 40 degrees
IVO: 21 degrees
IVC: 59 degrees
EVO: 61 degrees
EVC: 19 degrees
Estimated output: 221 horsepower at 5,000 rpm and 274 ft-lb torque at 3,000 rpm

Camshaft number: 85; Grind: "266"

Application: street
Advertised duration: 266 degrees intake and exhaust
Lobe separation angle: 109 degrees
Net valve lift: 0.300-inch intake and 0.299-inch intake
Valve lash (hot): 0.012-inch intake and 0.013-inch exhaust
Overlap (with advertised duration): 48 degrees
IVO: 24 degrees
IVC: 62 degrees
EVO: 62 degrees
EVC: 24 degrees
Estimated output: 223 horsepower at 5,000 rpm and 264 ft-lb torque at 3,500 rpm

Camshaft number: 85; Grind: "272"

Application: street and competition
Advertised duration: 272 degrees intake and 266 degrees exhaust
Lobe separation angle: 109 degrees
Net valve lift: 0.300-inch intake and 0.299-inch exhaust
Valve lash (hot): 0.012-inch intake and 0.013-inch exhaust
Overlap (with advertised duration): 51 degrees
IVO: 27 degrees
IVC: 65 degrees
EVO: 62 degrees
EVC: 24 degrees
Estimated output: 221 horsepower at 5,000 rpm and 257 ft-lb torque at 3,500 rpm

Manufacturer: Clay Smith Engineering

Clay Smith Engineering was established in the late 1940s and is still in business today (Buena Park, California). The company is well known for its logo, which features the cigar-chomping "Mr. Horsepower" woodpecker. Although some of the camshafts listed here have the same grind number as the Smith & Jones camshafts, there is a difference.

Camshaft number: Semi; Grind: "252"

Application: street
Advertised duration: 252 degrees intake and exhaust
Lobe separation angle: 110.5 degrees
Net valve lift: 0.315-inch intake and 0.314-inch exhaust
Valve lash (hot): 0.013-inch intake and 0.014-inch exhaust
Overlap (with advertised duration): 31 degrees
IVO: 16 degrees
IVC: 56 degrees
EVO: 57 degrees
EVC: 15 degrees
Estimated output: 216 horsepower at 4,500 rpm and 283 ft-lb torque at 3,000 rpm

Camshaft number: 3/4 Race; Grind: "260"

Application: street
Advertised duration: 260 degrees intake and exhaust
Lobe separation angle: 111.5 degrees
Net valve lift: 0.300-inch intake and 0.299-inch exhaust
Valve lash (hot): 0.012-inch intake and 0.013-inch exhaust
Overlap (with advertised duration): 37 degrees
IVO: 19 degrees
IVC: 61 degrees

EVO: 62 degrees
EVC: 18 degrees
Estimated output: 220 horsepower at 5,000 rpm and 272 ft-lb torque at 3,000 rpm

Camshaft number: Full Race; Grind: "266"

Application: street
Advertised duration: 266 degrees intake and exhaust
Lobe separation angle: 109.5 degrees
Net valve lift: 0.300-inch intake and 0.299-inch exhaust
Valve lash (hot): 0.012-inch intake and 0.013-inch exhaust
Overlap (with advertised duration): 47 degrees
IVO: 24 degrees
IVC: 62 degrees
EVO: 63 degrees
EVC: 23 degrees
Estimated output: 221 horsepower at 5,000 rpm and 265 ft-lb torque at 3,500 rpm

Camshaft number: Super; Grind: "272-2"

Application: street and competition
Advertised duration: 272 degrees intake and 266 degrees exhaust
Lobe separation angle: 111 degrees
Net valve lift: 0.313-inch intake and 0.299-inch exhaust
Valve lash (hot): 0.012-inch intake and 0.013-inch exhaust
Overlap (with advertised duration): 47 degrees
IVO: 24 degrees
IVC: 68 degrees
EVO: 63 degrees
EVC: 23 degrees
Estimated output: 224 horsepower at 5,500 rpm and 256 ft-lb torque at 3,500 rpm

Camshaft number: Super; Grind: "282"

Application: competition
Advertised duration: 282 degrees intake and 266 degrees exhaust
Lobe separation angle: 110 degrees
Net valve lift: 0.313-inch intake and 0.299-inch exhaust
Valve lash (hot): 0.012-inch intake and 0.013-inch exhaust
Overlap (with advertised duration): 54 degrees
IVO: 29 degrees
IVC: 73 degrees
EVO: 61 degrees
EVC: 25 degrees
Estimated output: 221 horsepower at 5,500 rpm and 242 ft-lb torque at 4,000 rpm

Camshaft number: Super; Grind: "282-2"

Application: competition
Advertised duration: 282 degrees intake and 272 degrees exhaust
Lobe separation angle: 111 degrees
Net valve lift: 0.313-inch intake and 0.312-inch exhaust
Valve lash (hot): 0.012-inch intake and 0.013-inch exhaust
Overlap (with advertised duration): 55 degrees
IVO: 29 degrees
IVC: 73 degrees
EVO: 66 degrees
EVC: 26 degrees
Estimated output: 222 horsepower at 5,500 rpm and 242 ft-lb torque at 4,000 rpm

Camshaft number: Super; Grind: "284"

Application: competition
Advertised duration: 284 degrees intake and 272 degrees exhaust
Lobe separation angle: 111 degrees
Net valve lift: 0.338-inch intake and 0.312-inch exhaust
Valve lash (hot): 0.012-inch intake and 0.013-inch exhaust
Overlap (with advertised duration): 56 degrees
IVO: 30 degrees
IVC: 74 degrees
EVO: 66 degrees
EVC: 26 degrees
Estimated output: 225 horsepower at 5,500 rpm and 241 ft-lb torque at 4,000 rpm

Camshaft number: Super; Grind: "284-2"

Application: competition
Advertised duration: 284 degrees intake and 282 degrees exhaust
Lobe separation angle: 111.5 degrees
Net valve lift: 0.338-inch intake and 0.312-inch exhaust
Valve lash (hot): 0.012-inch intake and 0.013-inch exhaust
Overlap (with advertised duration): 60 degrees
IVO: 30 degrees
IVC: 74 degrees
EVO: 72 degrees
EVC: 30 degrees
Estimated output: 221 horsepower at 5,500 rpm and 237 ft-lb torque at 4,000 rpm

Manufacturer: Weber Cam Grinding Company

Harry Weber founded Weber Cam Grinding Company in Los Angeles, California, in the early 1950s.

Camshaft number: n/a; Grind: "Track"

Application: street and competition
Advertised duration: 260 degrees intake and exhaust
Lobe separation angle: 113 degrees
Net valve lift: 0.317-inch intake and 0.315-inch exhaust
Valve lash (hot): 0.013-inch intake and 0.015-inch exhaust
Overlap (with advertised duration): 34 degrees
IVO: 18 degrees
IVC: 62 degrees
EVO: 64 degrees
EVC: 16 degrees
Estimated output: 220 horsepower at 5,000 rpm and 272 ft-lb torque at 3,500 rpm

Camshaft number: n/a; Grind: "F-1"

Application: street
Advertised duration: 264 degrees intake and exhaust
Lobe separation angle: 112 degrees
Net valve lift: 0.300-inch intake and 0.299-inch exhaust
Valve lash (hot): 0.012-inch intake and 0.013-inch exhaust
Overlap (with advertised duration): 40 degrees
IVO: 22 degrees
IVC: 62 degrees
EVO: 66 degrees
EVC: 18 degrees
Estimated output: 219 horsepower at 5,000 rpm and 269 ft-lb torque at 3,500 rpm

Camshaft number: n/a; Grind: "F-2"

Application: street and competition
Advertised duration: 275 degrees intake and 280 degrees exhaust
Lobe separation angle: 108.8 degrees
Net valve lift: 0.307-inch intake and 0.306-inch exhaust
Valve lash (hot): 0.013-inch intake and 0.014-inch exhaust
Overlap (with advertised duration): 60 degrees
IVO: 30 degrees
IVC: 65 degrees
EVO: 70 degrees
EVC: 30 degrees
Estimated output: 218 horsepower at 5,500 rpm and 248 ft-lb torque at 3,500 rpm

Camshaft number: n/a; Grind: "F-3"

Application: competition
Advertised duration: 300 degrees intake and 302 degrees exhaust
Lobe separation angle: 110.5 degrees
Net valve lift: 0.307-inch intake and 0.306-inch exhaust
Valve lash (hot): 0.013-inch intake and 0.014-inch exhaust
Overlap (with advertised duration): 80 degrees
IVO: 42 degrees
IVC: 78 degrees
EVO: 84 degrees
EVC: 38 degrees
Estimated output: 187 horsepower at 5,500 rpm and 199 ft-lb torque at 4,500 rpm

Camshaft number: n/a; Grind: "F-4B"

Application: competition
Advertised duration: 283 degrees intake and 282 degrees exhaust
Lobe separation angle: 111.3 degrees
Net valve lift: 0.301-inch intake and 0.299-inch exhaust
Valve lash (hot): 0.014-inch intake and 0.016-inch exhaust
Overlap (with advertised duration): 60 degrees
IVO: 32 degrees
IVC: 71 degrees
EVO: 74 degrees
EVC: 28 degrees
Estimated output: 214 horsepower at 5,500 rpm and 238 ft-lb torque at 4,000 rpm

Camshaft number: n/a; Grind: "F-4A"

Application: competition
Advertised duration: 283 degrees intake and 282 degrees exhaust
Lobe separation angle: 111.3 degrees
Net valve lift: 0.316-inch intake and 0.314-inch exhaust
Valve lash (hot): 0.014-inch intake and 0.016-inch exhaust
Overlap (with advertised duration): 60 degrees
IVO: 32 degrees
IVC: 71 degrees
EVO: 74 degrees
EVC: 28 degrees
Estimated output: 218 horsepower at 5,500 rpm and 240 ft-lb torque at 4,000 rpm

Camshaft number: n/a; Grind: "F-4"

Application: competition
Advertised duration: 283 degrees intake and 282 degrees exhaust
Lobe separation angle: 111.3 degrees
Net valve lift: 0.326-inch intake and 0.324-inch exhaust
Valve lash (hot): 0.014-inch intake and 0.016-inch exhaust
Overlap (with advertised duration): 60 degrees
IVO: 32 degrees
IVC: 71 degrees
EVO: 74 degrees
EVC: 28 degrees
Estimated output: 221 horsepower at 5,500 rpm and 241 ft-lb torque at 4,000 rpm

Manufacturer: Edward A. "Big Ed" Winfield

Edward A. "Big Ed" Winfield was a brilliant automotive engineer, considered by many to be the original hot rod camshaft grinder, and he designed the famous Winfield carburetor used on many racing cars in the early 1930s. Refer to Chapter 5 for more information about Big Ed. As a point of historical interest, Harvey J. Crane (founder of Crane Cams) credits Big Ed with the term "3/4 race," which referred to a camshaft ground by Big Ed with a performance level midway between his "semi" and "full race" grinds.

Camshaft number: n/a; Grind: "Semi"

Application: street
Advertised duration: 252 degrees intake and exhaust
Lobe separation angle: 112 degrees
Net valve lift: 0.238-inch intake and 0.281-inch exhaust
Valve lash (hot): 0.012-inch intake and 0.014-inch exhaust
Overlap (with advertised duration): 28 degrees
IVO: 19 degrees
IVC: 53 degrees
EVO: 63 degrees
EVC: 9 degrees
Estimated output: 198 horsepower at 4,500 rpm and 282 ft-lb torque at 3,000 rpm

Camshaft number: 3/4 Race; Grind: "3/4"

Application: street
Advertised duration: 260 degrees intake and 252 degrees exhaust
Lobe separation angle: 112 degrees
Net valve lift: 0.293-inch intake and 0.281-inch exhaust
Valve lash (hot): 0.012-inch intake and 0.014-inch exhaust
Overlap (with advertised duration): 32 degrees
IVO: 23 degrees
IVC: 57 degrees
EVO: 63 degrees
EVC: 9 degrees
Estimated output: 201 horsepower at 4,500 rpm and 276 ft-lb torque at 3,000 rpm

Camshaft number: Full Race; Grind: "R1"

Application: street
Advertised duration: 260 degrees intake and exhaust
Lobe separation angle: 112 degrees
Net valve lift: 0.293-inch intake and 0.291-inch exhaust
Valve lash (hot): 0.012-inch intake and 0.014-inch exhaust
Overlap (with advertised duration): 36 degrees
IVO: 23 degrees
IVC: 57 degrees
EVO: 67 degrees
EVC: 13 degrees
Estimated output: 210 horsepower at 4,500 rpm and 278 ft-lb torque at 3,000 rpm

Camshaft number: Super 1R; Grind: "SU-1R"

Application: street
Advertised duration: 268 degrees intake and 260 degrees exhaust
Lobe separation angle: 112 degrees
Net valve lift: 0.303-inch intake and 0.014-inch exhaust
Valve lash (hot): 0.012-inch intake and 0.014-inch exhaust
Overlap (with advertised duration): 40 degrees
IVO: 27 degrees
IVC: 61 degrees
EVO: 67 degrees
EVC: 13 degrees
Estimated output: 211 horsepower at 4,500 rpm and 269 ft-lb torque at 3,500 rpm

Camshaft number: Super 1; Grind: "SU-1"

Application: street
Advertised duration: 272 degrees intake and 268 degrees exhaust
Lobe separation angle: 112 degrees
Net valve lift: 0.313-inch intake and 0.301-inch exhaust
Valve lash (hot): 0.012-inch intake and 0.014-inch exhaust

Overlap (with advertised duration): 46 degrees
IVO: 29 degrees
IVC: 63 degrees
EVO: 71 degrees
EVC: 17 degrees
Estimated output: 220 horsepower at 5,000 rpm and 265 ft-lb torque at 3,500 rpm

Camshaft number: Super 1A; Grind: "SU-1A"

Application: street
Advertised duration: 280 degrees intake and 264 degrees exhaust
Lobe separation angle: 112 degrees
Net valve lift: 0.338-inch intake and 0.316-inch exhaust
Valve lash (hot): 0.012-inch intake and 0.014-inch exhaust
Overlap (with advertised duration): 50 degrees
IVO: 33 degrees
IVC: 67 degrees
EVO: 71 degrees
EVC: 17 degrees
Estimated output: 222 horsepower at 5,000 rpm and 257 ft-lb torque at 3,500 rpm

INDEX

MOTORBOOKS WORKSHOP

The Best Tools for the Job.

Other Great Books in this Series

Performance Welding Handbook
2nd Edition
0-7603-2172-8 • 139436AP

How To Paint Flames
0-7603-1824-7 • 137414AP

How To Build
Vintage Hot Rod V-8 Engines
0-7603-2084-5 • 138703AP

Honda & Acura
Performance Handbook
2nd Edition
0-7603-1780-1 • 137410AP

Hot Rod
Horsepower Handbook
0-7603-1814-X • 137220AP

How To Build the Cars of
The Fast and the Furious
0-7603-2077-2 • 138696AP

How To Tune and Modify Engine Management Systems
0-7603-1582-5 • 136272AP

Corvette Performance
Projects 1968–1982
0-7603-1754-2 • 137230AP

Custom Pickup Handbook
0-7603-2180-9 • 139348AP

Circle Track Chassis
& Suspension Handbook
0-7603-1859-X • 138626AP

How To Build A West Coast
Chopper Kit Bike
0-7603-1872-7 • 137253

101 Harley-Davidson Twin-Cam
Performance Projects
0-7603-1639-2 • 136265AP

101 Harley-Davidson
Performance Projects
0-7603-0370-3 • 127165AP

How To Custom Paint Your Motorcycle
0-7603-2033-0 • 138639AP

101 Sportbike Performance Projects
0-7603-1331-8 • 135742AP

Motorcycle Fuel Injection Handbook
0-7603-1635-X • 136172AP

ATV Projects: Get the Most Out
of Your All-Terrain Vehicle
0-7603-2058-6 • 138677AP

Four Wheeler
Chassis & Suspension Handbook
0-7603-1815-8 • 137235

Ultimate Boat
Maintenance Projects
0-7603-1696-1 • 137240AP

Motocross & Off-Road
Performance Handbook
3rd Edition
0-7603-1975-8 • 137408AP

How To Restore Your
Wooden Runabout
0-7603-1100-5 • 135107AP

Ultimate Garage Handbook
0-7603-1640-6 • 137389AP

How To Restore John Deere
Two-Cylinder Tractors
0-7603-0979-5 • 134861AP

How To Restore Your Farm Tractor
2nd Edition
0-7603-1782-8 • 137246AP

Mustang 5.0
Performance Projects
0-7603-1545-0 • 137245AP

For a FREE catalog visit WWW.MOTORBOOKS.COM or call 800-826-6600